P9-CDZ-836

The Grid

THE
GRID

THE FRAYING WIRES BETWEEN AMERICANS AND OUR ENERGY FUTURE

Gretchen Bakke

BLOOMSBURY

NEW YORK · LONDON · OXFORD · NEW DELHI · SYDNEY

Bloomsbury USA
An imprint of Bloomsbury Publishing Plc

1385 Broadway	50 Bedford Square
New York	London
NY 10018	WC1B 3DP
USA	UK

www.bloomsbury.com

BLOOMSBURY and the Diana logo are trademarks of Bloomsbury Publishing Plc

First published 2016

© Gretchen Bakke, 2016

ISBN: HB: 978-1-60819-610-4
          ePub: 978-1-62040-124-8

LIBRARY OF CONGRESS CATALOGING-IN-PUBLICATION DATA

Names: Bakke, Gretchen Anna, author.
Title: The grid : the fraying wires between Americans and our energy future / Gretchen Bakke.
Description: New York : Bloomsbury USA, 2016.
Identifiers: LCCN 2016001376 | ISBN 9781608196104 (hardback)
Subjects: LCSH: Electric power systems—Technological innovations—United States. | Electric power distribution—United States—History. | Clean Energy—United States. | Electric power failures—United States. | Energy Policy—Social aspects—United States. | BISAC: TECHNOLOGY & ENGINEERING / Power Resources / Electrical. | SOCIAL SCIENCE / Sociology / General. | BUSINESS & ECONOMICS / Infrastructure. | TECHNOLOGY & ENGINEERING / Environmental / General. | SCIENCE / Electricity.
Classification: LCC TK1005 .B27 2016 | DDC 333.793/2—dc23 LC record available at http://lccn.loc.gov/2016001376

2   4   6   8   10   9   7   5   3   1

Typeset by RefineCatch Limited, Bungay, Suffolk
Printed and bound in the U.S.A. by Berryville Graphics, Berryville, Virginia

To find out more about our authors and books visit www.bloomsbury.com. Here you will find extracts, author interviews, details of forthcoming events and the option to sign up for our newsletters.

To Guillaume, for whom books are written.

Any sufficiently advanced technology is indistinguishable
from magic.
—ARTHUR C. CLARKE

Right now there's three power companies in New York City:
there is ConEd in Manhattan, there is the Brooklyn power
company Brooklyn Union Gas up in Brooklyn, and there is a
windmill here on 519 East Eleventh Street.
—INTERVIEW FROM THE 1978 FILM *VIVA LOISAIDA*

Contents

Introduction   xi

CHAPTER 1
The Way of the Wind   1

CHAPTER 2
How the Grid Got Its Wires   25

CHAPTER 3
The Consolidation of Power   57

CHAPTER 4
The Cardigan Path   85

CHAPTER 5
Things Fall Apart   115

CHAPTER 6
Two Birds, One Stone   149

CHAPTER 7

A Tale of Two Storms   185

CHAPTER 8

In Search of the Holy Grail   219

CHAPTER 9

American Zeitgeist   255

Afterword: Contemplating Death in the Afternoon   291
Acknowledgments   293
Notes   295
Index   343

Introduction

Energy is a hot-button issue these days. From the marble halls of state to Louisiana's once-battered and besmeared coastline, we as a nation keep hearing about the need to make the transition to green, clean, sustainable energy. Doing so, we are told, will help put our jobless back to work, contribute to America's triumphant reemergence on the world stage, and quite possibly save the planet. At the very least, it will spare us from the oil spills, rig explosions, fracking quakes, and mining disasters that come without warning and destroy without compunction. So, too, we hope that more sustainable forms of power might help guarantee us a future with fewer superstorms and snowpocalypses, rising tides and shrinking ice caps. The stakes for changing how we make and use energy have never been higher.

Transitioning to sustainability is the big, long-term challenge. But left out of this picture is the fact that for the most part, America does not run on gas, oil, or coal any more than we may one day run on wind, solar, or tidal power. America runs on electricity. The age of information has remade the ways we communicate, socialize, and even learn things into electricity-dependent processes. And though we tend to think of this shift in terms of computerization, the computer is only the tool. Electricity is what makes it work. As computers become ever more essential to how we live, travel, and care for ourselves, so, too, have electric appliances brought us bright, functional homes and workplaces. Our health and our hospitals are intensely dependent upon electric power, as are our factories, ports, police, and military. Even money these days is electrically made and stored data that we sometimes convert into pieces of paper and the jingle of coins. Some of us even smoke electric cigarettes, and it's looking likely that soon we'll be shuttling about in self-driving electric cars. Electricity has become so essential that using the word "blackout" to refer to a power outage is something of a misnomer. Losing light is the least of our problems when our electricity system now crashes to ground, which it's doing with increasing frequency.

Our energy sources might be the main cause of our worry—coal with its carbon emissions, natural gas with its methane emissions and questionable extraction methods, nuclear with its poisoned waste, and even oil with its cost in wartime blood—but they are not all that matters to the story of American power. The rest of the tale begins at the precise moment these fuels go up in flames. We don't use them raw. And while lots of oil still goes into cars, almost all the others are used to power the grid—a complex and expansive electrical delivery system that we care little for and think even less about.

Though it is situated squarely at the center of our modern lives, for most Americans the grid rarely breaches consciousness. A couple of years ago I was shown a snapshot a friend had taken of a sunset, its dusky pinks and evening grays bisected straight through by power lines; it even had a couple of utility poles smack-dab in the middle of the frame. "Look," she said, "what a beautiful sunset." The grid was right there and it didn't even disturb the view. It's the world's largest machine and the twentieth century's greatest engineering achievement and we are remarkably oblivious to it. Experts at unseeing its wires and poles, we are equally unlikely to be able to differentiate a power station from an oil refinery. Substations, essential to our grid, flit perhaps at the edges of recognition, though most have been well hidden behind cement walls or stuck in out-of-the-way corners of our cities and towns. Transformers, once a world-altering technology, have also been rendered utterly unspectacular, secreted inside gray canisters that cluster like coconuts at the tops of urban utility poles. Even our smart phones, which also form a part of the grid, albeit one that can be disaggregated and carried around in a pocket, don't seem properly infrastructural. Yet open one up and what you will see is, in effect, some circuit boards carefully wrapped around the mass of a battery. Its power source is its heart, and this is only fed by a constant return to the grid.

As such, the grid would seem to best Frankenstein in the heterogeneity of its parts, to outdo the Eisenhower Interstate System in its geographic scope, and to dwarf space exploration in the complexity of its science. It also reveals a bizarrely accurate picture of this nation in the particularities of its historical unfolding. The grid is not just something we built, but something that grew with America, changed as our values changed, and gained its form as we developed as a nation. It is a machine, an infrastructure, a cultural artifact, a set of business practices, and an ecology. Its tendrils touch us all.

Literally speaking, our grid is comprised of the battery, its port, the charger, its plug, the outlet and its wall-hidden wires out to the street, the transformer and the familiar forest of urban utility poles, the low-voltage wires that run along these poles out to the substation, more transformers (bigger here), some syncrophasers, relays, switches, and fuses and the giant pylons that march away across the empty parts of America carrying high-voltage wires to the power plant, where it all begins in an electromagnetic generator spun fast by a steam-heated, or wind-blown, or water-wheeled, or gas-combusted turbine. Our grid is a peculiarly pervasive infrastructure that touches every life, pierces every wall, bifurcates every landscape, and runs every battery.

And though we call it *the* grid, in America we actually have three of them: one for the West that includes a tiny bit of Mexico and much of western Canada; one for all of the East; and a separate, smaller one for Texas. For the most part Mexico has its own grid, but Canada does not (except for its errant province, Quebec, which, like Texas, has chosen to keep its options for secession open by managing its infrastructure for itself).

Looking at its sleek lines and high towers, one might mistake the grid for an electrical transportation superhighway. But to those intimately acquainted with it, another image springs to mind—that of an old, beat-up, pothole-riddled, one-lane dirt road. The grid is worn down, it's patched up, and every hoped-for improvement is expensive and bureaucratically bemired.

More than 70 percent of the grid's transmission lines and transformers are twenty-five years old; add nine years to that and you have the average age of an American power plant. According to the industry expert Peter Asmus, we rely on twice as many power plants as we actually need because of "the massive inefficiencies built into this system." As a result, significant power outages are climbing year by year, from 15 in 2001 to 78 in 2007 to 307 in 2011. America has the highest number of outage minutes of any developed nation—coming in at about six hours per year, not including blackouts caused by extreme weather or other "acts of God," of which there were 679 between 2003 and 2012. Compare this with Korea at 16 outage minutes a year, Italy at 51 minutes, Germany at 15, and Japan at 11. Not only do we have more outages than most other industrial countries, but ours are getting longer. The average U.S. power outage is 120 minutes and growing, while in the rest of the industrialized world it's less than ten minutes and shrinking. According to Massoud Amin, a power systems engineer, on "any given day in the U.S. about half a million people are without power for two or more hours."

For every minute of every one of these blackouts, money is lost and national security is at risk. Our economy literally droops when the voltage on our wires sags (brownouts), and it seriously stutters when the power breaks for real. Big outages are the most dramatic in this regard, but they are not necessarily the most expensive. The 2003 East Coast blackout, caused by an overgrown tree and a computer bug, blacked out eight states and 50 million people for two days. So thorough and vast was this cascading blackout that it shows as a visible dip on America's GDP for that year. Six billion dollars lost: that's $60,000 per hour per blacked-out business of lost revenues across 93,000 square miles.

As impressive as this was, outages of five minutes or less are actually more costly to the efficient running of our national economy, in part because they are so much more prevalent. For a lot of machines, an outage of fifteen seconds and one of fifteen minutes or fifteen hours produces precisely the same kind of damage and takes roughly the same time to put right again. After the big East Coast blackout of 1965, caused by an incorrectly set relay, we couldn't even get our power plants to come back online. It turns out you need electricity to make electricity, and so diesel generators had to be wheeled in and used to "black-start" our coal-burning plants and nuclear power stations.

Another biggie for New York City came in 1977, caused by a lightning strike at a substation up the Hudson River. As this one tore the lid off Pandora's box, it released another kind of trouble. The looting was tremendous. The blackout gave the anger long brewing on that city's streets an outlet. New York rang with the sound of smashed glass and stank of soot and smoke as scores of tenements were burned to the ground. There are many ways that darkness can be destructive to the well-being of a community. It can bring cold and hunger, as did the ice storms that blacked out the Ozarks in 1987. Equally, it can play host to hot tempers and opportunistic predators.

On a lighter, if no less economically devastating, note in 1987, and again in 1994, the NASDAQ lost power because of squirrels gnawing on the electric lines that connect that stock exchange to the world. Not every blackout is caused by a world-class storm. Many are made by wildlife, squirrels most especially, and even more originate with trees—so many that overgrown foliage is the number one cause of power outages in America in the twenty-first century.

If the grid's only problems were that it is old, worn down, and vulnerable to attacks by maleficent forces as diverse as kudzu and commandos (people do occasionally go after the grid with shotguns), we might just shore it up a bit, pay it slightly more mind, and continue on our way. Just keeping the trees trimmed, the storm damage to a minimum, the supply chains in motion, and the squirrels away from the wires is plenty of work. Add in worries about terrorist hackers that have come with modes of grid modernization that use computing to "smarten" our electrical delivery system, and clearly we have an epochal infrastructural upgrade on our hands.

There is, however, another issue pressing its way into the system that brings a new urgency to the inevitable task of reforming our grid. It turns out that transitioning America away from a reliance on fossil fuels and toward more sustainable energy solutions will be possible only with a serious reimagination of our grid. The more we invest in "green" energy, the more fragile our grid becomes.

A coal-burning plant might be bad for the environment, it might be bad for the miners who struggle underground to bring that coal up and bad for the West Virginia mountains razed in its production, but it's a remarkably good fit for the grid. And not just coal. The grid is at its best when we make electricity using what are called "stock resources." These are fuels that when we use them, we use them up—plutonium, natural gas, oil, coal, and anything else dug up from the ground or grown upon the earth that we burn and then have no more. These steady fires are what our twentieth-century grid was built for. During that period, the business of making and delivering power was a remarkably centralized activity, run by "natural" monopolies (the utilities) that built infra-structure according to a top-down system of command and control. Renewables, especially the two most popular, solar and wind, have been creeping their way into our national energy mix since the 1970s. These are anything but obedient. The utilities can command them, but they do not listen. The utilities can attempt to control them, but their efforts falter in the face of meteorological processes, planetary in scale.

At issue are the vagaries of nature. The wind can never blow with the same steadiness that factory-combusted coal provides, and the sun's rays, while ever present, are all too often blocked from reaching the panels we've built to gather them by clouds. Each moment of shade, no matter its length, is translated by solar panels into a dip in the electricity they produce. Every shift in the wind's speed becomes an uneven rise and fall of available electrical current.

This is further complicated by the fact that our machines, which strictly speaking are what the grid was built to power, were designed to run on a well-regulated, predictable electrical flow. For over a century this steadiness and predictability was made at the power plant. Sustainable energy sources provide something else: an inconsistent, variable power that our grid is unprepared to adapt to. Nevertheless, it is increasingly clear that these alternatives to fossil fuels are where our energy future lies.

If greening the grid were only a matter of transitioning to variable generation, I suspect that the grumbling and foot dragging might be assuaged by some well-thought-out policy decisions and lots of seed money thrown at the development of some new and better means for storing electricity (see chapter 8). However, it is problematic not only how renewables make power but *where* they make it. Wind farms go up where it is windy. And places like Wyoming or Iowa or West Texas have a lot of strong wind on constant offer. What they don't have are many people to use this power or very good long-distance power lines to carry it to more promising markets. The grid was never built to be robust in the midst of wastelands. But these empty, often uninhabitable places tend to be where wind and solar power are most effectively produced.

No less problematic for the grid is the phenomenal recent growth in home solar systems, which as of this writing are being slapped up pretty much everywhere it's sunny (and even in some places it's not). These are, in essence, tiny renewable-power plants. We all use the electricity made by panel owners in aggregate, in exactly the same way we use power generated by a natural gas combustion turbine or a hydroelectric dam. When we pay our electric bill, we are also now paying the producers of this homemade electricity for the power they contribute to the whole.

In theory this is all well and good, except that in America we have long made electricity in huge quantities in big, centralized power plants. Our grid was built with these factories at its heart. Everything about the way it is structured, from the merest wire to the biggest sub-station, was designed for the effective transit of power from a few massive producers to a wide scatter of users. All of us everywhere use electricity, but we never used to make it. Home solar installations turn this logic on its head without doing much to help reconfigure the grid, which in certain pockets of our nation now has to take power from everywhere and distribute it to everywhere else.

To apply another sharp tack of misanthropy to the ebullient balloon of green-energy optimism, transforming our grid for the twenty-first century does not just mean integrating newer technologies with the old. The grid isn't just some contraption wired together out of various bits of this century and bits of the last (plus a substantial shake of nineteenth-century ways of doing and building as well). It's also a massive cultural system. And the stakeholders—the utilities, investment firms, power plant owners, mining firms, and "too-big-to-fail" multinational conglomerates—will not go gently into the future's bright night. An abundance of carbon-free energy is a nice idea, but fossil fuel companies are still responsible for the vast majority (66.5 percent) of the power on our grid. Coal may be on the way out; the Energy Information Administration (EIA) predicts that roughly a fifth of total coal capacity will be retired between 2012 and 2020, and even the chief executive of the American Coal Council admits that the industry has abandoned any plans to replace the retiring fleet.

The recent boom in natural gas, however, seems to legitimate that industry's claim that it is the perfect transition fuel. For the moment, it's almost as cheap as coal; we have a lot of it, ripe for domestic exploitation; and it is more efficient for producing electricity because power plants that run on natural gas are spun first by the raw force of combustion and then run on steam, giving twice the bang for the same buck. At first glance, natural gas is also cleaner than coal. This is certainly true if one is thinking in terms of particulates and not greenhouse gases or wastewater pollution. Methane emissions from natural gas development and leaky transport infrastructure, however, undercut some of the climate advantage gas has over coal. One 2012 study estimated that if the industry were to let leak only 3.2 percent of the gas produced, it could be worse for the climate than coal. In most places, leakages are not that bad. In some places they are worse.

Even nuclear power, which is problematic on many levels, has gained a new sparkle as "green" has come to mean carbon-free rather than pollution-free. They can still melt down, and the problem of storing the radioactive waste these plants produce has never been adequately solved. It's hard to figure out how China, for example, can transition away from a reliance on coal without a significant uptick in nuclear power development. The cost of these plants means that anywhere with a market economy has almost entirely stopped building them, but the technology is being improved. There are a number of ongoing efforts, globally, to scale nuclear technology way down, making it portable and mass-producible. If successful we may be surprised to find nuclear blooming back up on our soil one future day.

Existing ways of making power for our grid involve much more than the raw fuels we choose between. They are also massive networks of business interests, geo-political stakeholders, and carefully wrought legal structures. These took decades, and in some cases centuries, to build, and each has a specific inertia that will need to be shifted, or in some cases overcome, in the process of remaking the ways and means of American power.

The grid, then, is built as much from law as from steel, it runs as much on investment strategies as on coal, it produces profits as much as free electrons. Even fair-minded politicians make poor decisions as they navigate the labyrinths of legislative change. For-profit power companies, deeply subsidized oil and natural gas interests, the mining industry, and even the railroads all have vested interests in keeping America dependent upon fossil fuels. These are big, unwieldy corporate machines. They do change with time, but it is a glacial kind of change characterized by recalcitrance and torpor. Nor should we be surprised that these behemoths of old power are unwilling to embrace a transformation that will quite likely put them out of business. In this they are no different from any company—though it is easy to valorize smaller more innovative endeavors—none of them commit suicide on purpose. Maneuvering to capture market share is a big part of what is driving the energy transition, and it is in a constant state of vacillation. One step forward, two steps back, we lurch rather than cruise into the future. The changes to our grid will not happen overnight, but they are already far enough along to be rightly considered irreversible.

As the first decade of the twenty-first century crested, making power from renewable sources shifted from a nice idea and a minor player on the electricity scene into the mainstream. The speed and the scale of this shift have been truly extraordinary—a momentum that is only likely to increase. It is estimated that by 2050 "nearly every single power plant in the U.S. will need to be replaced by new plants." If, then, in

2015 we had three times as much wind power as in 2008 and twenty times as much solar, these looming plant retirements only further open the door to renewable technologies, big and small. In Hawaii, at present, over 12 percent of the houses are equipped with solar panels—so many that on certain sunny days these home solar systems produce more electricity than the state needs.

In the summer of 2015, Hawaii's utility began refusing further connections by home solar owners to the grid. Not because they are mean-spirited, but because they can't use it all, and excess power on the grid causes bits of it to shut down self-protectively—a measure that spares the infrastructure but blacks out its users. The same was happening in Vermont, which is saying something, because Hawaii might have a lovely climate for solar power, but Vermont is both northern and gray. Nevertheless, its two largest utilities had, for much of 2015, also called for a hiatus on new grid connections for home solar systems as these were causing the bills of their customers without home solar to rise precipitously. Until something infrastructural, in some cases, and fiscal in others must be done to deal with all the homemade electricity being pumped into the grid during daytime hours on sunny days, both states will be sticking with a baseline of fossil fuels despite the fact that both would openly and honestly prefer a greener route.

This reveals another curious thing about our grid: though it may be big, it is also intensely local. For example, right now, variable generation is producing only a tiny portion of the electricity used in this country, about 7 percent overall. But in some places this number is much higher. In Texas, wind power alone accounts for 9 percent of electricity production, in Oregon it's almost 13 percent, and in Iowa it's a stunning 30 percent. There is so much wind in some places in the United States right now that on particularly blustery days, the local balancing authority—charged with making sure the amount of electricity going into the grid and the amount being drawn from it are exactly the same—has to pay some of the wind farms to shut down their turbines and also pay large industrial concerns to take and use more power than they actually need. In Texas, one blustery September day in 2015, the price per megawatt-hour of electricity dropped to negative 64¢. The utilities were actually paying their customers to use power. Everything has gone a bit topsy-turvy.

By 2025 the Department of Energy wants 25 percent of America's electricity to come from renewables. It sounds like a pipe dream, until you consider that between 2009 and 2014 the amount of renewable power traveling the lines of our grid has more than doubled. And this federal goal is actually rather lightweight when compared to certain state efforts. Maine is aiming for 40 percent by 2017, California for 50 percent by 2030—and these numbers don't even include the electricity made from rooftop solar systems. Vermont, ever the tiny optimist, has a goal of 75 percent by 2032. Hawaii is aiming for 100 percent. These are not impossible objectives, but they will require us to utterly reimagine our grid.

At issue is that when these goals are set, the grid rarely enters the larger conversation. At times it is almost as if it doesn't exist. We all—you and I, our state and federal legislators, our business leaders and upstart entrepreneurs, even the world climate change experts—we just dream our dreams, run our coffee machines, take our snapshots of fine sunsets, and set our exaggerated goals for the future without ever seeing that between us and everything we envision stands this technological monument to recalcitrance. The grid is there. It's time we learned to notice it so that we, hand in hand with its momentous complexity, can step forward into a brighter day, a brighter night, and a cooler, less storm-muddled planetary future.

Though we tend think about the grid in ways borrowed from the physics of water pipes that hold, transport, and deliver drinking water or those of natural gas fuel lines that bring methane to stovetops and hot water heaters, this is not in the least how electricity works. The lines are not hollow with electricity inside them, they are solid through and through. Nor does electricity flow, trickle, or drip. It is not a liquid or a gas subject to the laws of fluid dynamics; it's a force. We make it—which is already pretty awesome—by breaking electrons free from their atoms (at the power plant) and then allowing these to bump into their next nearest atomic neighbor, dislodging their electrons, and allowing these to bump along to the next. Some metals, at the atomic level, make this process of dislodging electrons from atoms easy, and these are the metals we use to build conductors (power lines). The lines, in turn, play material host to this atomic domino effect that moves at roughly the speed of light. Though this tumble of electrons is the fastest thing in our known world, actual atomic drift is only about .05 inches per second, or roughly the speed of cold honey. To human eyes, all this action at the atomic level looks like static, cool metal, but this is the falsest of views. The electricity we make and use every day is effectively inseparable from the wires of the grid. And yet the product chains we've designed to make (and make money off of) electricity, ignore the ways in which electricity and its infrastructure are different from all other kinds of products. This has become even more the case since legislation passed in the early 1990s effectively made electrons and, say, bananas identical objects from the market's point of view. Never mind that electricity has so little in common with a banana that the two might as well have originated in different physical universes—we now treat and trade them in an almost identical way.

We can touch a banana, box it, even let it sit awhile if the price is not right for sale. We can engage people to grow a banana and then not buy it from them. We can charge a specific price for a banana that customers can know and judge before they purchase it, and if they find one brand, variety, or color of banana less appealing than another, they can choose to buy a different one. Preferences matter to supply and demand, and the possibility of choosing from a limited selection of bananas has become part of what defines the product and also what makes it work—and make sense—within our economic system.

Electricity is not like this, ever. It cannot be boxed or stored or shipped. It is always used the same instant it is made, even if the person using it is a thousand miles from the source. If electricity is made, it's shipped; if it's shipped, it's used. And all of this happens in the same singular millisecond.

For our part, consumers, who know very well what a banana is and what a dollar is and how the two kinds of things stand in relationship to each other, rarely have any real sense of what a kilowatt-hour is, how much it should cost, how many we are using, and how much we might be paying for them. To complicate matters further, the utilities are essentially state-supported monopolies, making the electricity business like a store where everyone within a certain area is obliged to shop. At this store no prices are listed on individual items (these unlisted prices are also subject to change without notice), and at the end of the month, shoppers are issued a single bill for the entire month's worth of purchases. Without specific information, one study showed, consumers of electricity "have a hard time estimating the costs and benefits of their actions." This is part of why we pay electricity no mind. Understanding it means understanding the grid, and both function according to terms beyond the scope of daily life.

Yet because the electricity we depend so heavily upon cannot be separated from its infrastructure, we cannot reform our energy system without also transforming our grid. There can be no electric power without all the machines and wires that make it, make it safe, and translate it across our great nation into our lightbulbs, toasters, and air-conditioning machines. We also have to pay for this system, something that renewables have complicated as they mix up producers with consumers in entirely new ways. And, because it's America, the whole shebang needs to convey a profit into somebody's pocket. Almost all the big utilities are investor-owned, which means they have shareholders who have been promised at least the occasional dividend. Corners get cut in order to ensure that this flow of cash continues apace, and decisions get made with profit motives in mind that brook little concern for the particular capacities, and incapacities, of the grid. We may imagine the grid as primarily a machine to make and move electricity, but integral from the very start was that it also make and move vast quantities of money. A lot of people are still happy with this way of doing things.

The grid's current shape has a lot to do with the specifics of its technological and business history. No stranger to chaos, in its earliest days America's electric grid grew haphazardly, not radiating out from one imaginary center point but spreading instead like a pox, appearing only in spots with dense enough populations to ensure a profit. This is why for its first half century, electricity was largely an urban phenomenon. That finally changed during the Great Depression, when the government intervened and brought the grid and electricity to the rural folks whom capitalism would have happily left behind.

Despite the absence of an overarching plan and only occasional intervention on the part of state and federal governments, the grid quickly grew in both size and complexity. And though it has been upgraded periodically as technological developments have leapt forward, these upgrades have been evolutionary, not revolutionary; every improvement has had to mesh with the existing system. No advance could be made in one place that couldn't accommodate weaknesses somewhere else down the line. Little by little, this build here, that reform there, developed into the grid we have: it is a jury-rigged result—highly inventive in places, totally stodgy in some, fantastically Rube Goldberg in yet others. As implausible as it must sound, the machine that holds the whole of our modern life in place "works in practice, but not in theory." No one can see, grasp, or plan for the whole of it.

Wrapped up in this is how the utilities established themselves as monopolies and saddled the industry with the long-term inertia a lack of competition too often begets. In part because of its protected status as a regulated monopoly, the grid worked reasonably well from the moment of near to universal electrification in the late 1930s through the end of the 1960s. But when it stopped being true that utility profits were assured by an ever-increasing rise in demand, and as their fundamental strategy of grow-and-build simultaneously faltered, the grid began to slip into a state of disrepair. The utilities, which for so long had thrived on being the least inventive, least flexible, most run-of-the-mill companies in America, didn't know how to change. This indifference, even from within the industry, when coupled with legislative attempts to make the electricity business more profitable as the digital economy rose to prevalence in the final years of the twentieth century, have made the grid an ever more uneven technology. The parts that can be leveraged to make money get a lot of attention while the rest silently molder.

Before the blackouts began in earnest, apart from a few concerned engineers and environmentalists, among them Amory and Hunter Lovins, not many people cared much about the consequences of this protracted inattention. In 1979, in the wake of America's second energy crisis that decade, the Lovinses prepared a report for the Pentagon on the state of America's domestic energy infrastructure. Their conclusion was surprising: national security was threatened more by the "brittleness" of America's electrical grid than by possible future disruptions in the flow of imported oil. As prophetic as that report would turn out to be, few were really listening, apart from Jimmy Carter, who would lose his presidential power soon after its publication. And so years passed.

Decades passed. And time transformed the system the Lovinses had labeled weak and brittle into an even more fragile version of itself.

And then the inevitable happened.

In the early years of the new millennium, the grid finally began to break; its business model cracked open as much as its copper and cement were worn down. California suffered blackouts so severe the governor declared a state of emergency and one of its major utilities filed for bankruptcy (the first for a large utility since the Depression). Then a nuclear power plant in Vermont toppled over. It literally fell down; its support structure had rotted right through. Not because it hadn't been subject to regular maintenance and inspection, but because all the cameras set to watch the infrastructure age were pointed elsewhere and the maintenance checklists didn't include the beams that held the cooling tower up. Bits of the Pacific Northwest's corner of the grid started to crash not two or three times a winter—as had long been the case—but two or three times a month. The local storms have a whole new kind of blow to them. Texas watched helplessly as its grid went down again and again, the blame falling on excessive air-conditioner usage. The White House even lost power, not once but twice during the Bush years (and twice since!).

The Executive Branch, in 2013, finally admitted that "grid resilience is increasingly important as climate change increases the frequency and intensity of severe weather," promising that Americans can expect "more severe hurricanes, winter storms, heat waves, floods and other extreme weather events being among the changes in climate induced by anthropogenic emissions of greenhouse gasses."

Big storms, and other weather weirdnesses, have ironically become one of the main drivers of contemporary grid reform, but in the least efficient way imaginable. We use fossil fuels, including natural gas, to make electricity, the chemical pollution from these contributes massively to global warming, global warming makes for more ferocious storms, and these storms swoop in and decimate the grid.

This destruction prompts people to think about ways that the grid might be made harder to destroy. Occasionally they take action on these thoughts and some small element of our infrastructure gets changed. If our aim is to prompt infrastructural reform that will work for us over the long term, then this absurdist loop is clearly the least efficient and most destructive route possible. And yet it is the one we take.

Our grid, its form, and the way this fits into modern life is holding us back from our energy future—not just from the introduction of sustainable sources of electrical generation but also from meaningful conservation and, most important, from energy independence. By this I do not mean just that we might be free from the vagaries of Middle Eastern oil (or British drilling companies or our own mining conglomerates), though this, too, has its benefits, but also that we stand on the brink of a new kind of energy independence: the type that comes from being able to make our own energy decisions as individuals, households, neighborhoods, or towns. The scale of our infrastructure is changing and people, in the process, matter more. Almost anything that can be subdivided is being cut up into smaller and smaller parts. Individuals, small businesses, towns, and counties are developing the means to stamp their feet, vote with their pocketbooks, and be heard. This was already true yesterday and it will be even more true tomorrow.

The U.S. military, for one, is unable to tolerate even the tiniest of voltage fluctuations (these make computers behave as if they have been possessed by demons, and they are common on America's grid) or risk physical attacks on unsecured public infrastructure. As a result the military is in the process of converting all their domestic bases to microgrids. Google is doing the same for its headquarters and data centers, as are the states of Connecticut and New York for their towns. New York alone is in the process of constructing eighty-three new microgrids. Others, including CitiBank, *Businessweek*, and the

Edison Electric Institute, are predicting the end of the electrical utility company as we know it. Perhaps they will remain stewards of the wires, but no longer will they make power (they already mostly don't) nor will they make money from selling it.

Of all the myriad "unintended consequences" and "creative reactions" arising from ruined bits of our current system, the most important is this: inventors and corporations, every bit as much as individuals, are thinking about what can be accomplished by small actions and collective efforts. They are doing so en masse for the first time since the domestication of electricity in the 1880s. The cumulative effect of this new emphasis on "small" is to wrest control away from big men accustomed to fighting their battles on a grand scale. America's infrastructure is being colonized by a new logic: little, flexible, fast, adaptive, local—the polar opposite of the way things have been up until now.

This kind of thinking, and the activities that follow from it, confound the powers that be. If you want to understand just how much, think about the extent to which the U.S. military is flummoxed when beset by smaller, more mobile, more local, more flexible, more adaptive, and more creative militant forces. This is what it feels like to be a utility company, a regulatory agency, or a governmental body in the electricity game right now. The ways in which they managed American power since the end of the nineteenth century are slowly but surely being relegated to the trash heap of history. Sometimes the people leading this assault look like Silicon Valley smart-guys, sometimes they look like aging hippies finally getting their way, sometimes they look like multinational corporations, and sometimes, every so often, they look like a retired schoolteacher in seasonally decorated knitwear.

Together these are the foot soldiers on the attack. The object of their ire is not so much the big grid but the habitual and increasingly ineffective and uninteresting ways that the electricity game has been run. And they, in their diversity and their passion, allow us to witness new logics emerging: small systems and solutions are gaining in currency over big ones, flexible or variable ways of doing things gain favor over rigid ones, mobility and portability are more appealing than the static and the fixed, and wirelessness is championed whenever and wherever possible. Initially it will seem chaotic, and it will be chaotic, as a thousand interested actors grab for a piece of what has long been a proprietary infrastructure. But within this mess, certain unspoken cultural attitudes will proliferate, and from these, patterns will emerge.

This book brings all of this into focus—the ruins and the dreams, together with the incomparable complexity of our grid's technology; its history, replete with absurdities and brilliances, together with the people, laws, and logics that brought it into being—so that as historical and technological exigencies press down upon us, we, the users of the grid, might understand the stakes and implications of our choices and its failings a little better. The grid might look stable, its presence as steadfast as it ever was; it might feel known, its electrical power almost as reliable as it ever was. But we would do well to wise up to the fact that both impressions belie an intense seething change in the very structure of the power machine that keeps us all warm, lit, and, relatively speaking, well off.

CHAPTER 1

The Way of the Wind

D ay one. It's a bright autumn morning in Washington, D.C. I and about four thousand other people, most in business suits, have already made it through four tight rings of security, descending at each stop-and-check farther underground until we pass the final metal detector and emerge into a startlingly well-appointed, underground bunker of a conference center, the Ronald Reagan Building and International Trade Center. The lighting is subtle, the decor an elegant symphony of beige. For the next five days this venue will play host to the (mostly) men who spend their lives making, regulating and transporting electricity to American homes and businesses. Welcome to Grid Week.

This is the human side of the grid, not its wires and poles, substations and power plants, but industry executives, electrical engineers, and utility-company representatives, some of whom are newly in the business of smartening the electric grid; some even are small-tech entrepreneurs. All of them play a part in making it work. We are here this early-autumn morning to hear the conference's first keynote address, to be given by Stephen Chu, a Nobel Prize–winning physicist and also for a time the U.S. secretary of energy. As we settle into our seats, Dr. Chu steps up to his place at the podium and the auditorium falls respectfully still. He, too, has a quietness to him, a demeanor that with his delicate frame and slight baldness make him look more like a monk than a bureaucrat. In an odd way, I think, perhaps he is both. This keynote will be both a sermon and a policy speech.

What Dr. Chu is going to tell them, the men who keep our grid up and running, is to integrate more renewable power generation—more wind, more sun, more waves and tides, more geothermal, more of everything that is hot without being heated and that moves without being pushed—but first he is going to tell them some horror stories.

Sure enough, by about slide 5 (after we have learned many great things about our energy future and how America will soon be rocketing back to unprecedented international success), things start to look really bad.

"On September fourth, 2008," says Secretary Chu, gesturing with his laser pointer at a massive PowerPoint display, "at just before five P.M. in Alamosa County, Colorado, a thick layer of clouds swept across the sky." He pauses to glance down at his audience.

No one is coughing or shifting in their seat, nothing beeps or buzzes. We are attentive.

In fact, given how the room feels, I imagine myself in a cluster of ten-year-old boys with ratty sneakers listening to ghost stories around a late-summer campfire instead of with several thousand middle-aged industry men in well-pressed pants. The other difference is that rather than the play of shadows at the edges of a fire's circle of light, we are given statistics and sharp lines on graphs. This one, the one Chu is pointing at now, plummets precipitously downward.

"Five minutes later," he continues, "there was a jagged but rapid eighty-one-percent drop in the electricity output from the solar farm that served the community."

Eighty-one percent. Five o'clock P.M. A nicely drawn downward-plunging line.

Everyone in the room knows exactly what is going on; what they don't know is how to deal with it. An all but instantaneous 81 percent drop in generation at five in the evening when everyone is coming home from work, switching on their air-conditioning, TV sets, and computers is the kind of story that sets the hearts of electrical engineers palpitating. Electricity consumption on this, the world's largest machine, must at every moment be balanced with electricity production. The more solar there is in any given mix of "fuels" used to generate electricity, the harder it is to cope with the sudden arrival of a cloud, especially at five in the afternoon when things on the demand side have just shot through the roof. On the graph, the black line labeled GENERATION is pointing straight downward while the red line labeled CONSUMPTION is angling up and up toward the sky. When solar is how you generate your electricity, no sun means no power. It puts these very men, in their workaday lives, into a ferocious scramble to avoid a blackout.

"Four months later," says Secretary Chu, continuing blithely on to the next slide, "on January fifth, 2009, in the Columbia River Gorge, the wind stopped blowing quite suddenly and didn't start again for three weeks." He pauses again. He lets the ramification of this massive and long-term stoppage wend its way through his audience. Three weeks. *Three full weeks.* You could have heard a pin drop.

"Meanwhile, all twenty-five of the Gorge's wind farms lay still."

No wind means no generation, and no generation means no power. Yet all the people that live downline from these farms, many of them lefty Northwesterners who believe strongly in the integration of renewable resources like wind and solar into electricity production, aren't just going to sit around happily for three weeks without any electricity. Even the greenest of consumers aren't going to just wait for the wind to pick up again before checking their e-mail or making some toast. Even if just a portion of their electricity comes from wind power, someone somewhere is having to make up for this calm, an adjustment that still, in most cases, involves firing up some other massive power-making machine. It's not impossible, but it's a struggle: it's hard to do well, harder to do fast, and almost impossible to do cleanly. America's backup power plants are the oldest and dirtiest in the fleet. They should have been decommissioned and torn down decades ago. Instead we use them as a last-ditch resource when power supplies fall short. We use them a lot.

It's not just that machines have to respond to the variability of renewables. It's also that the culture of electricity making has to be transformed. The power plants forced to take up the slack when renewables fall still are matched pretty well in age with the people in charge of running them. As a later speaker in the day's proceedings would point out, 60 percent of men who run our electricity system are within five years of retirement. A quick glance around and I would have to agree. The people in this very room are at the end of their careers. They, together with the institutions they work for, have long had one way of doing things, and now they are scrambling to adapt to a changed landscape. Before grid-scale wind and solar power came online, slow and steady always won the race. There was no competition in electricity, a protection enshrined in law that made each utility the unique master of its realm. They made our power and they always knew how much of it there would be, where it would come from, and where it would be used. Plans were made seasonally, collegially, as every four months utility men would sit down in a room and talk about how the winter might go, or the spring, or the summer; these men made sure there were enough power plants chugging along to provide what they estimated to be the right amount of electricity. Except for the occasional panic of a too-hot day or too-cold one—when demand for electricity jumps precipitously—their plans pretty much worked.

Nothing in the system they grew up in, and now run, prepared them for a means of power generation that not only varies from minute to minute, but which they do not own, cannot control, and have no plan for. The new world of privately or corporately owned variable generation, strewn about every which where, demands that they be very light on their feet. But the utilities, the utility men, and 2,500-megawatt (MW) coal-burning power plants don't dance much. It's an industry that plods along and likes it that way.

Their balletic capacities to the side, the utility companies do find themselves trapped in an increasingly tight spot between a rock (variable generation) and a hard place (keeping the lights on). If in 2009, when Chu gave this presentation, there were twenty-five wind farms in the Columbia River Gorge; today there are four times that many, most containing hundreds of turbines, each turbine producing well over a thousand kilowatts of power. Some of the largest wind developments in the nation sit nestled into this single slash of land. All that power, currently estimated at 6,000 megawatts (or enough electricity to power 4.5 million households) depends solely on the way the wind blows.

And when the wind, ever fickle, stops its blowing all the electrons these vast machines have been built to harvest out of thin air disappear. It's that simple. The grid must be balanced; consumption must always match production, for there is as of yet no real means of storing that electricity for later use. If power is not being made right now, somewhere, somehow, we simply don't have it to use.

As impossible as it may seem to people outside the industry, grid-scale electricity storage hardly exists. There are some artificial lakes pump-filled with water that folks in mountain states can call on in a pinch, but that's about as far as it goes. For now, no household has a cookie jar full of watts secreted away for later use; no nation has a strategic electricity reserve. As a result the electricity we use, day in and day out, is always fresh. So fresh, that less than a minute ago, if you live in wind farm territory, that electricity was a fast-moving gust of air. And if you live in coal country, it was a blast of pulverized coal dust being blown into a "firebox"—a huge, industrial, flash-combusting furnace. If you live in hydro country it was a waiting rush of water dammed up by a massive concrete wall. Picture it. The electricity you are using right now was, about a second ago, a drop of water.

But it's not water anymore. We like to talk about electricity as "flowing" from one place to another, as if we could predict where it might go once we've released it onto the wires. But we can't. It doesn't flow downhill, it doesn't take the shortest path, nor will it follow one route at the expense of another. The wires we use to transport electric current from where it is made to where it is used aren't much like pipes, or mains (as they are often called). Nor can electricity really even be said to "flow" through them. Wires are conductors, which is to say that they are metal, and to the extent that something electric happens because of them, it seems to happen as much outside as inside the lines. Power lines are there to channel or direct broad halos of electromagnetism in a direction determined by something as simple as someone depressing the lever on their toaster. Suddenly a pathway opens up, one that wasn't there an instant before, and electricity follows it, moving into and through the toaster, where it is slowed down as it passes. This slowing down, or resistance, produced by the device causes electrons to release heat, which toasts the bread. After a certain number of seconds the lever pops back up, ejecting the toast and closing off the toaster channel, and electricity must find another way.

It's not a system that needs to be planned. No one decides which electrons will go to Los Angeles to make doughnuts and which to Walla Walla to make toast; all of the electrons are going everywhere at once. As long as there is a "sink" (of the kind caused by a toaster toasting), all the electricity on the grid will move toward it by whatever means possible. The reason your toaster doesn't explode every time you turn it on is because there are thousands, indeed millions, of other sinks on our grid where other devices are making the same kind of "hey, no resistance over here" calls to the available electric current.

There are also a million little devices on the grid, and some big ones, to standardize the voltage, or push, of that electricity, so that the power available to your toaster in the first place is substantially less than that traveling along high-voltage lines from the dam to the nearest substation. Your average outlet is already offering the toaster access to one of the mildest intensities of electricity available. You still don't want to pry a burnt nub of bread out of there with a butter knife—a nasty shock greets that activity—but were you to so much as touch a low-voltage downed residential wire, it would kill you. The system, in its current form, is designed not only to protect the toaster, but to protect us from the potential force that even the modest voltage of domestic electric current delivers.

Toasters don't explode, wires function well, lightbulbs go on when the wall switch is flipped, all because the grid is kept in balance: there is enough electricity available to run our machines, but there is not so much that it rips through and destroys them.

This is our grid in a nutshell: it is a complex just-in-time system for making, and almost instantaneously delivering, a standardized electrical current everywhere at once. And though schematas of the grid tend to make it seem like there is a line out of a power plant that ends in the toaster, the whole thing is actually a giant loop that both starts and ends at the power plant, or generating station. These factories make an electric current by tearing electrons out of their atomic orbit and then give them no real choice but to power the whole system as they make their way, rather quickly all things considered, back into these orbits again. The power plants that accomplish this electron-ripping task can be made to run on wind, or natural gas, or coal, or uranium hexafluoride, or dried cow dung; any fuel will do. Strung into this loop are the manufactures, businesses, farms, and toasters that use the power electrons release as they pass by. Whatever power needs these consumers and their things have at any given instant in time has to be balanced pretty much perfectly with the amount of power being produced at that same instant way on down the line. This is as true of a customer who turns on their porch light as it is of one who brings a new server farm online. This is why peak load— when customers suddenly use a lot more electricity than they were using just five minutes before—is a startling kind of problem for utilities. It's also why figuring out ways to design our world to use power when it is made, rather than whenever we feel like it, is a brain-twisting, but fundamentally smart, idea.

Variable generation—the technical term for power plants that make electricity out of unpredictable fuel sources like the wind, sun, or waves—is a problem. It doesn't matter which end of the system escapes control. It can be us, using too much power all of a sudden (like when we all come home after a long day's work and simultaneously turn up our air-conditioning just as the wind slackens), or it can be cloud cover, stripping the generative capacity from solar panels. Regardless, the utilities and other balancing authorities have to act very quickly to set things right again. Otherwise there just isn't enough power in the lines to keep the lights on. Lots of blackouts start this way.

The rub is that, with the exception of hydroelectric dams, the output of all existing, comfortable-to-utilities, means for generating electricity take significant time to turn up or down. The wind can stop instantaneously. A cloud can blow over the sun just as quickly. Or, ten thousand customers can turn on their air conditioners. When this happens a controller sitting in front of a wall of flat screen monitors in a control room somewhere sees it: *bam*, an 81 percent drop in output or an 81 percent increase in demand. It's a precipitous curve graphed on a screen; it's a red warning light blinking in mechanical panic; it's a buzzer irritating in its insistence; it's a nerve-jangling phone ringing and ringing and ringing. Someone on the other end needs a fix and he needs it now.

I sat in the control room one day for the company that runs 54 of America's wind farms scattered from the Gorge to the Arizona desert to northern New York State and I saw how it worked. Response time is limited by human and mechanical capabilities, but wind speed and lightning storms are not. There are predictive mechanisms in place. The man in front of those screens sees lightning strikes move their way across a map drawing closer and closer to repair crews until he picks up the phone and makes the call: "Get out of there, shut it down." The same is true of the wind, usually. The weather is predicted to be blustery or calm. The spinning of the turbines is closely monitored, and electricity is priced for sale based upon expected output.

But sometimes a lull comes, like the one Chu has pointed to, and it's total, stable, and unexpected. At this instant there is nothing the controller can do, there is no dial at his right hand that he can just turn to increase the output of some other generating plant on the same lines. Not so quickly at least. He will turn dials and push buttons and make calls. He'll do whatever he can, but the physics of electrical generation from "stock" or man-made resources such as coal, natural gas, or nuclear are against him. And though they are getting faster, they are just not very adjustable. Coal-burning plants, at 50 percent in five minutes, are one of the fastest; natural gas (from a cold start) takes about ten minutes to get up to speed, while nuclear takes a full twenty-four hours to turn up, though it can be shut down in seconds.

In human time, five minutes might seem pretty quick, given that we are talking about moving a mechanical system as massive and complicated as a coal-burning power plant, which pulverizes and combusts, on average, 125 tons of coal every five minutes.

But in electricity time, which is what matters to grid stability, five minutes might as well be infinity. In five minutes, electrical current generated by a power plant outside Muncie, Indiana, can go to Mars. Even in the decidedly imperfect conditions of electrical transmission more characteristic of life on Earth, the wind power generated in the Columbia River Gorge that is not used by the relatively sparsely populated states of Oregon, Washington, and Idaho can be easily transported along a long DC (direct current) line to the good people of Los Angeles County, where it is gobbled up by air conditioners well before its sixty seconds are up. This is one of the reasons the grid is big. Big means that power plants can be built in places with not too many people but still provide electricity to large population centers as distant from one another as Seattle is from San Diego.

At times, however, even the grid's remarkable span is insufficient to absorb all the power produced in the Gorge. For this river valley is not only a phenomenal source of wind power, as Secretary Chu pointed out, but it has an extensive hydroelectric infrastructure left over from the heady days of big government investment in public works that helped to pull America out of the Great Depression. These New Deal dams (Grand Coulee and Bonneville most especially) and their smaller, more recent brethren were providing 98 percent of the Pacific Northwest's electricity needs before the first industrial wind turbine went up; now there is all that hydro and all that wind power all in one place.

Washington, Oregon, and Idaho, they could live bright, warm, electric lives without the wind. The rain, snow, and meltwater are more than sufficient. In fact, of all the power produced in the Gorge, from whatever source, only about 15 percent is used locally. The rest is shipped on down the lines to whoever will buy it. This is why it is a big deal when the wind stops blowing for three weeks. It's not just some widely scattered Left Coast ranchers that lose power but it's also city people and townsfolk all over the Western United States and certain choice bits of Canada as well.

The fact that we don't yet have a good means of storing electricity doesn't just mean that we have little backup power on hand to deal with shortages; it also means that it is difficult to dispose of surplus power when it's produced in excess. While our dependence on oil may have taught us to think about and prepare for interruptions in supply, it has never happened that instead of a carefully measured tank of gas, what you get at the filling station is a giant splash of the stuff tumbling down and over and utterly inundating you and your car. But solar and wind power see to it that the energetic equivalent of this great messy slosh is happening to the grid all the time. Anywhere in the nation with a high concentration of wind turbines or a high concentration of photovoltaics always runs the risk of generating more electricity than can be easily consumed. This is the part of the renewable energy horror story that Secretary Chu left out.

You can't just turn the wind down. When it blows hard, those turbines spin and spin and the output is tremendous. The young control room operator with whom I sat watching the weather as it approached and moved through widely scattered wind farms told me with a note of awe in his voice that you can actually see a gust of wind as it tops the Rockies and then hits one set of turbines after another all the way to the coast. You can see it in the power spikes—*bang, bang, bang*—of wind farm after wind farm shooting electricity into the system. It floods the grid; it crashes through the infrastructure much like a wave crashing against a sea wall on a stormy day. Even Los Angeles can't absorb all the electricity made on a seriously blustery day in the Pacific Northwest. Even the Western Doughnut, as the high-voltage DC line that carries electricity from the Gorge to the people of Southern California is called, with its 3,100 megawatts of transmission capacity (or half of L.A.'s peak capacity), cannot carry it all.

When there is too much power on the wires they overload, or circuits break to protect them, and in so doing they close, rather than open, available paths for excess power to take. It's hyperbole that your toaster will explode; the system will self-protectively black itself out long before your toaster turns into a bomb of flame on your kitchen counter. In this way, blackouts should be seen as source of grace as much as a bane and a burden.

Imagine, then, that Secretary Chu's story does not stop with the wind's unpredictable calms, but rather continues to include its more impetuous, tempestuous side. Imagine that instead of moving on to a discussion of solutions he follows his harrowing tales of what unexpected cloud cover means for solar output and what a lengthy calm means for wind power with a third story of the relationship between renewable energy and our electric grid, for this story is also true.

"On the afternoon of May nineteenth, 2010," he might have said, "in a single chaotic hour, more than a thousand wind turbines in the Columbia River Gorge went from spinning lazily in the breeze to full throttle as a storm rolled out of the East." Here he would pause, to see if his audience understood what was about to happen, what all of this wind was about to do to all those turbines. "Suddenly, almost two nuclear plants' worth of extra power was sizzling down the line—the largest hourly spike in wind power the Northwest has ever experienced."

A massive uncontrollable, unmanageable, unstorable, undumpable electricity surplus. Chaos on the lines. And what is worse: it's May.

In Oregon in May it's still raining. It's been raining since November, and it will continue to rain for another month or so before things begin to lighten up. In the Cascades, the mountain range that bifurcates the state, all that rain is snow, and in May all that snow is meltwater—pure, chill runoff. The rivers are very full, and they are sloshing their way down the sides of mountains and hills into the man-made lakes that sit behind, and feed, each and every dam on the mighty Columbia. In May these reservoirs can't hold another drop. They are full up. And the turbines on every dam up and down that mighty river chug along at a fearsome rate, because if they don't there are only two options. Either the reservoirs flood up over the homesteads, highways, and towns that dot the river's edge, or the dam operators let the water out through spillways.

Though the second might sound like a good option, sadly for them, it also happens to be illegal. Because, in May, the fishlings are running, tiny silver slivers that will, in two to three years, grow into beautiful, fleshy oceanic salmon. If the dams flood their spillways, these fingerlings will be ravaged. Their numbers will be decimated year by year, and not only will the commercial salmon industry be threatened, but the species itself will slip slowly from plentiful to endangered, from dinner plates to Grandma's memory bin.

So, spilling the water isn't an option, at least not in May.

The only option is to let the dams operate at close to maximum capacity. And if the dams are going to make all the power they can, they are going to need all available transmission lines to move that power out of Oregon to anyone and everyone who looks like a market. It can't be stored, it must be transported and used immediately, or the land will flood, or the grid will crash. This is every day in May. There is water, there are fish, there are laws, there are power lines with a finite capacity to transport electricity, and there is a market that just might not be big enough to use all the power they are being fed.

Power production isn't just an industry, it's an ecology. And renewable resources are not just about the planetary good kept from public offer by corporations with other visions for their own profitable futures. Making American power is about how technological, biological, and cultural systems work in concert to keep our lights on, our basements and roadways clear of flood water, and fresh fish on our tables. It's delicate in all sorts of ways. Though I will concentrate largely on infrastructural delicacy in this book, it does the reader well to remember that the vulnerability of the grid as a technological system is intimately linked to the fragility of biological systems (like salmon runs), the intractability of legal and bureaucratic systems (like the endangered species act), and the unpredictably of meteorological systems (like wind storms).

The wide-scale integration of variable forms of power generation didn't create this situation. The grid's entanglements with culture and law and natural systems were always there. Renewables have just made these entanglements impossible to ignore; they stress the existing system just enough that all the delicate balances reached over the passage of a century are thrown off-kilter. As all of these diverse bits of what make our grid work interlock and entangle, there just isn't a lot of room for quick action. Once people with politics and profit motives get their fingers into the briar patch, it seems at times like there is no room to act at all.

This was the situation into which the equivalent output of two nuclear power plants was suddenly poured that mid-May day back in 2010. The only real option was to shut down the wind turbines. Switch the beasts off. Still their spinning. Clear the lines. Let the storm blow itself out. Leave all those electrons unreaped.

At the time, that's exactly what the local balancing authority—the Bonneville Power Administration (BPA)—mandated be done. They called up the corporation that developed, built, and still manages most of the wind farms in the Gorge, the Spanish-owned conglomerate Iberdrola, and asked them to pretty please, and yes, immediately, turn off their many hundreds of wind machines, whipping around just then at absolutely ferocious speeds in the onslaught of wild air.

But what does Iberdrola care for the grid? They are in the business of making electricity, not of moving it to market. Transmission is the utilities' problem and balancing is the balancing authorities' problem, regulation is the regulators' problem, interregional cooperation is the ISO's problem. Iberdrola's problem, as the second-largest wind company in the world, is maintaining a profitable bottom line. Turning off their turbines at a moment of maximal productivity? Well, it's just not a sensible course of action. Most especially because the federal subsidies that have helped them to build and maintain their almost three thousand American-sited turbines only accrue if those machines are turned on and running. It's not just that they only make money from these beasts' ceaseless rotation, it's also that they have to *pay back* money if their turbines are ever off. Even the agency that Chu headed didn't imagine as it wrote up its guidelines for subsidies that sometimes the best thing anyone could do with a wind turbine is turn it off; that sometimes, in America, we can have too much of a good thing. If Iberdrola switches off even one turbine just to be nice, this has very real ramifications for their profitability. From their point of view, if the grid isn't up to the task of moving to market the power they make, then the grid needs to be better.

In this they are right. Every man at Grid Week knows it. That is part of why they have come.

Over and over, investments in renewable sources of power generation are failing or falling very short because America's electric grid just isn't robust enough or managed well enough to deal with the electricity these machines make. And not just in the Columbia River Gorge.

In West Texas, the largest wind farm ever planned on American soil was abandoned in 2008 because the utility refused to build a high-voltage line out to the site. And the developer, the local oilman T. Boone Pickens, thought it was a travesty given how much he was investing to build the farm itself that he would be expected to also build the transmission infrastructure. He shelved the project after having installed just a thousand turbines, a fraction of the total.

Add to this a second outrage. Pickens had already been obliged to use turbines that were small by international standards, just as was every other wind farm developer in America at the time. The grid's fragility demanded it. If a wind storm can turn a field of "small" wind machines into the equivalent of a nuclear power plant in a period of minutes, you can only imagine what would happen to a field of the really big ones. Germany's Enercon makes a 7.5 MW model (only slightly smaller than the largest offshore turbines, which come in at 8 MW), whereas in the United States the most common turbines remain the 1.5 MW GE model and the slightly bigger 2 MW Gamesa. This has nothing to do with how fast the wind blows across American plains versus German ones; it has everything to do with the wires these massive machines feed into. It is the system that stands between the point of generation and point of consumption that delimits productivity. The grid is the weakest link. It isn't made for modern power.

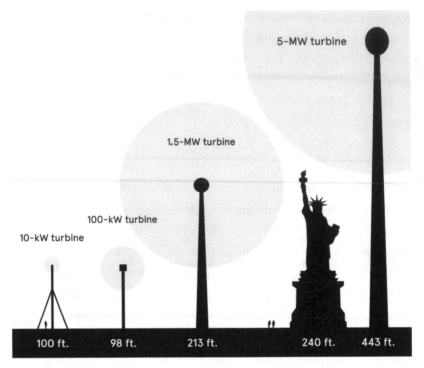

FIG 1 The height, swept area, and power rating of common wind turbines

So this one windy May day in the Gorge may have been exceptional as regards the quantity of electricity suddenly surging through the wires, but the problem of too much unpredictable electricity is all too ordinary, in the American West and Plains states especially.

In the spring of 2011, when I interviewed Elliot Mainzer, then BPA's director of strategic planning and now its acting director, the balancing authority had once again just paid the wind farms in the Gorge to shut down. This time not because of a storm, but because of an exceptionally robust runoff. The dams needed all the space on the wires. Mainzer, who is both a realist and an optimist (a rarely seen combination of traits), predicted then that "at the current rate of wind development the region's system of dams and power lines will start running into consistent operational problems around 2013, when wind in the agency's territory reaches total capacity of some 6,000 megawatts."

In other words, we've already passed the point of no return. If they were shutting down the wind farms with some regularity in 2010 and 2011, right now, with 6,000 MW of power rolling out of thin air and into our grid, at precisely this point, Mainzer predicted that the grid "will require major structural changes"—adding, after a pause, "If it's done right it's a huge opportunity."

This, then, was the substance of the fear felt in the soles of the feet of every well-shod industry man sitting in the Washington, D.C., auditorium that lovely September day listening to the secretary of energy speak about the problems of integrating renewables into the existing grid. They knew, to a man, that it would be better for them and for the reliability of the technology they are charged with shepherding into the twenty-first century if we could just stick with coal and natural gas, nuclear and hydro. The complexity of the rest, with its now-you-see-it, now-you-don't volubility, with its fiefdoms and awkward economics, with its ties to knotty physical systems, and with its unpredictability across domains—it's all a terrible headache to an industry whose job is to keep the lights on no matter the social, technological, fiduciary, meteorological, or political circumstances.

And then, after all of this, Secretary Chu smiles. He looks down upon them from his podium and drops the bomb they all knew would come.

"The Obama administration," he said, "has set a goal of 25 percent renewable energy use by power producers by 2025; ten percent by 2012." And before the men in attendance could leap to their feet and demand with one unmodulated voice: "Yes, but HOW!?" Secretary Chu continued unperturbed through the rest of his PowerPoint presentation, which amounted to a tidy list of solutions to the grid's known woes: use smart grid technologies, curb customer demand, end peak demand, develop grid-scale storage, add a nationwide extra-high-voltage DC/AC transmission network, reduce line congestion, encourage interregional cooperation, develop interoperability standards, increase government investment, train a new generation of grid operators, and integrate large numbers of electric vehicles.

This is the "solutions" laundry list, and a pretty thorough one, especially if we add deployable energy efficiency to the mix. Almost all of it lands squarely in the laps of the utilities and their regulators—sometimes friends, sometimes enemies, always themselves trying to balance investments with profits while maintaining infrastructural integrity. Because if they don't there is no state entity that can just step in and make the grid work if the utilities have to declare bankruptcy or otherwise fail at their appointed task. There is no backup system to the grid. If we can't make it work, then it doesn't. It's as simple as that.

If you listen carefully you will notice that Chu's laundry list is the same set of solutions, in part or in whole, that have become the talking points of anyone interested in reforming the grid for most of the last decade. But rattling off the list is not the same thing as getting the state of California, balky after the deregulation debacle of the late 1990s, to talk to anybody about cross-border transmission. Pointing out a series of best practices is not the same as persuading consumers to let their obtuse utility company take remote control of their home air-conditioning. And throwing money at the problem is not the same as figuring out how to get Vermonters (or anyone else) to allow high-voltage lines to be built in their backyards.

The complexity of the situation is way beyond anything actually captured by the talking points or anything resolvable by a repeated return to stated goals.

We know, with the benefit of hindsight, that the interim goals of the Obama administration's 2009 renewable energy plan have not exactly been met. Though the numbers do look surprisingly good on first glance. According to NREL (the National Renewable Energy Laboratory—a thirty-five-year-old federal institution that is something like the NASA of renewable energy), 12.4 percent of America's electricity was made from renewable resources in 2012. Read the small print, however, and it immediately becomes clear that slightly more than half (55 percent) of the total still comes from hydroelectric power.

Drought years to the side, the dams are steady. In 2000 they were generating about 78,000 megawatts, and in 2012 they were generating about 78,000 megawatts, though this should rise somewhat in the near future as the big old dams are "returbined"—their efficiency raised by the integration of newer technology. In most cases, however, when a "renewable energy" goal is issued by an administration, or anyone else, it is usually cast in terms of "nonhydro" renewables. And that number, for the United States as a whole in 2014, was 6.76 percent—though, because electrical production is still largely a local affair, it's much more than that in certain pockets, such as the Columbia River Gorge, California's Altamont Pass, Arizona's deserts, Hawaii, the Dakotas, Iowa, and West Texas.

This 6.76 percent renewable generation within our nation's still largely fossil-fuel-driven electrical economy might make wind and solar seem negligible in absolute terms. However, a closer look at the numbers reveals something remarkable, something that grid engineers already know: the recent growth in these two domains has been nothing short of explosive. In 2012, wind power installations accounted for 75 percent of all new generation in the United States, while installed solar, still a tiny piece of the electricity pie—only 0.3 percent of the on-grid electricity in the country—was nevertheless up by 83 percent over 2011 (and in 2011 it was up 86 percent over 2010). If in 2012, a banner year for American solar, only 30 MW of new concentrating solar power was brought online, then 2013 is nothing short of meteoric—with 900 MW planned. That's a 3,000 percent increase in a single year.

In 2015, the Obama administration virtually promised that these trends will continue upward for at least the next fifteen years, by legislating a 30 percent cut to 2005 CO_2 levels by 2030. The largest producers of CO_2 in the United States are coal-burning plants for making electricity, and the only way to meet these goals is to close hosts of them: "The ambitious rules hope to remake the nation's electricity system by closing hundreds of heavily polluting coal plants while rapidly expanding the use of natural gas plants, wind and solar power." In response to which, Nick Akins of American Electric Power, a Midwestern utility, responded with a simple threat: "If the proposed rule stands, there will be blackouts."

Nor do these "national trend" numbers include home installation of solar panels, which are contributing in their own way to the mounting crisis of infrastructural management. Though there are issues of excess and shortage tied into "net-metering"—when electric companies pay homeowners for the power their solar panels feed back into the common system—this crisis for utility companies is as much one of cash as it is of current. In certain expensive markets, like Hawaii and Southern California and in certain sunny ones, like Arizona and, recently, New Mexico, homemade solar power now costs about the same or even slightly less than grid-made power. Why, then, pay a utility company for something you can make for yourself? No good reason at all. Quite suddenly, the utilities aren't earning enough money to perform basic upkeep on the grid, though all of their customers are still using it. Solar-panel owners feed power into the grid during the day, but they draw electricity exclusively from the grid in the evening and at night. To cover basic infrastructural costs utilities in regions with a lot of rooftop solar are charging those customers without solar panels, the ones still getting all their power from the grid, higher rates. This of course leads these folks to switch to solar as well. The situation has gotten so bad in Hawaii that in 2015 the state's utility refused to enroll any more customers' net-metering programs. People can still put solar up on their garage roofs in Hawaii, but the utility won't connect to them, won't pay for the electricity they generate, and won't offer any kind of deal to homeowners on power consumed after dark. This cycle hadn't yet reached crisis level in 2009, when Chu was listing the known woes of power companies. It's at crisis level now.

What we are bearing witness to are the early days of a variable and distributed generation revolution. Electricity is being made everywhere, by power producers of all sorts and sizes, and increasingly from uncontrollable and largely unpredictable means. And because of an awkward piece of legislation called the Energy Policy Act (1992), which laid the foundation for the deregulation of the electricity industry, in many places not only have the utilities lost control of who makes power and how and where they make it, but they have also lost the right to own power plants themselves.

The Energy Policy Act separated electricity generation by law from electricity transmission and distribution (a divorce formalized by the Federal Energy Regulatory Commission's Order 888 issued in 1996). In effect this means that private companies can build condensed solar power plants wherever the sun shines hottest, individual home owners can mount solar panels on anything that doesn't move, and multinational conglomerates, or farmers, can install wind farms wherever the wind blows most ferociously—as well they should, for these are the sites that are most efficient when it come to the generation of electricity.

What is new with the Energy Policy Act is that these investors in electrical generation, large and small, don't need to give much thought as to how the grid, in often very out-of-the-way places, might deal with the influx of unpredictable power. Nor do they need to care for how utility companies will manage the task of keeping people's lights on when they are faced with the problem of too much power one instant, and too little the next. And even where the utilities do have a modicum of control over the stability of generation, they are losing control of their revenue streams, through rooftop solar, through the loss of big power plants, through the advent of real-time electricity markets, and through interventionist rate making by regulatory agencies that control how much customers will be charged for their electricity. It begins to seem that in the not too distant future the companies we now call "utilities" will become stewards of the wires and little more. But the wires, of course, are the only piece of the whole system that generates no revenue save a small rental fee to those who use them to pass electricity from one cash cow power plant to a thousand or a million paying customers.

For the moment this is mostly a problem in the West, on the part of the grid known as the Western Interconnection, and in Texas, which has its own grid. With numerous offshore wind farms planned in Lake Erie and in the Atlantic, off the Eastern Seaboard, the renewable power problem besting grid managers on the Western Interconnection is about to become an onus on everybody.

Renewables and their scattershot siting are not what make America's electricity difficult to manage in the second decade of the twenty-first century. They just brings to light a problem that has been characteristic of our grid for more than half a century: it was made to be managed according to a command and control structure. There was to be total monopolistic control on the supply side of great electric loop—which included generation, transmission, and distribution networks—and ever-increasing yet always-predictable consumption on the customer side of things. Electricity would move from one to the other, while cash would move in equal measure in the opposite direction.

This system was always partly fantasy, but it also mostly worked for a long time. Even the early big blackouts like the one that took down much of the Eastern Interconnection in the early 1960s were to be blamed more on systems complexity than on flaws in the logic undergirding the grid as a whole. Today it's a different story.

Every time America changes, whether a little bit or a lot, infrastructure lags behind. The things we build, especially the big things, and the institutions we invent to support these are far more permanent than the ways we choose to live. The 1950s were not the 1970s.

People lived in different parts of the county, they bought different products in different quantities, they consumed different amounts of power at different times of the day, they lived in different-sized houses, pursued different professions, and raised their children with different values. Yet the grid of the 1950s was in many ways the grid of the 1970s. And the grid of the 1970s was in many ways that of the 1990s. For the most part it is still our grid today.

When it comes to our electrical grid, decades pass, half centuries pass, and the logical structures that underlie its mass, most of its machinery, and most of the people educated to work on it age and are patched up, but they are rarely replaced. And then something happens to disturb the balance like the Energy Policy Act or the mass deployment of some really good wind turbines, or a 50 percent drop in the price of solar panels and the whole system reels. The grid, its values, and its base technologies have been out of true for decades, but renewable and distributed forms of power generation have pushed the whole system over the edge of the easily recoupable.

The grid will have to be reimagined, it will have to be reinvented, and parts of it will have to be rebuilt. This would have happened without the mass introduction of wind and solar power, but these have hastened the realization of the necessity of change. Or, to borrow the words of a recent article in the *Los Angeles Times*: "The problem is that renewable energy adds unprecedented levels of stress to a grid designed for the previous century."

It's worth considering in more detail what this previous century's grid actually is, where it came from, and why we have so long retained its most basic premises and components.

Our grid might have long been an inflexible, brittle, monopoly-managed monolith, but that is not how its story started out. In the beginning, electricity was a highly local affair. At times, in fact, it seemed we might not end up with a national grid at all, but rather a system of household-sized generation plants with no wires at all between buildings. Then for a while we also had a bunch of "microgrids" with a generation plant or two and a system of wires for a relatively speaking tiny "community" of users. Nowadays this is how quite a few college campuses, prisons, and military bases make their power, but back in the early days of electricity, unlike today, these designated grids all ran different voltages of electricity. They all also overlapped geographically. There was one voltage for streetcars, one for the lights, one for industrial concerns, and each of these had its own private system of wires. There were so many wires; the sky was a black spaghetti tangle of wires. In the 1800s electrical infrastructure was an absolute mess.

From these inauspicious beginnings we got a national grid with power plants far from view, long loping lines between us and them and, nearer at hand, distribution networks strung through neighborhoods, that link individual houses by means of pole-top transformers to the system as a whole. That this is how electricity works in America and pretty much everywhere else in the industrial world is not the logical outcome of physics, it's the product of cultural values, historical exigencies, governmental biases, and the big money dreams of financiers.

In order to understand why we have a grid at all—and why we have this one in particular—we need to jump back a bit, to the earliest days of electrification, and watch how the grid was invented and built into a brittle, inflexible machine of massive scope and unimaginable complexity that is nevertheless remarkably egalitarian. Our grid delivers electricity as easily to the poor as it does to the rich, and it blacks out privilege almost as often as poverty.

CHAPTER 2

How the Grid Got Its Wires

Electricity is not like anything else. It's not a solid, or a liquid, or a gas. It isn't quite like light or heat. It doesn't move like the wind or the tides. It doesn't combust like oil or burn like wood. If it resembles anything at all from the world we know, it is in some way like gravity. Which is to say, it is a force to be reckoned with.

Unlike gravity, electricity is lethal, most especially in its wild form—lightning—though the threat of electricity's killing side is always there, even when it's been most thoroughly domesticated. This is why we don't touch downed wires, don't do much home electrical repair ourselves, and discourage children from sticking bobby pins into outlets. In the early days of the grid, people didn't even change their own bulbs for fear of electrocution. Instead, a trained bulb replacer was dispatched on a bicycle balancing a giant sack of hand-blown vacuum-filled ampules on his back to replace all the bulbs that had burned out during the previous weeks.

Although we now know that electricity is a force and we understand in nuanced ways how it works—much of which will be explained in this chapter—what is curious about the grid's earliest days is that then we did not. Nobody knew what electricity was until long after Edison's first grid was built and had burned down. A circus elephant was killed by electricity, and then a man, and numerous dogs, as its precise killing force was tested without being properly understood. Electricity was harnessed, made, transported (all wired up), and caused to light and power any number of early devices while continuing to secret within its ineffable physics the cause of its capacity to "communicate life" to inanimate beings and, equally, its ability to steal this life force from animate ones.

Despite a lack of clarity on the specifics, what became clear as the technology for making electricity and transmitting it was finessed was that it was uniquely capable of powering things at a distance and doing so very close to instantaneously. Thus did it become possible to make electrical power "over here" and use it "over there" at essentially the same moment. Electricity's peculiar capacity to divorce space from time was in keeping with other nineteenth century inventions: the telegraph (1830s) sent "messages" across town and later around the world in something like an instant; the telephone (1876) did the same with the voice itself; so, too, the radio (1896) with its wireless communication of sounds and songs everywhere all at once; and the phonograph (1877) with its reproducible, if slower and more material wax and later shellac discs that allowed recordings made in one spot to be heard in quite another.

As each of these technologies gained ground in the popular imagination, and with it a certain share of the market, they displaced a less effective technology. The telegraph caused the post to seem painfully slow, while the telephone soon rendered the telegraph if not obsolete a notably awkward way of dealing with long-distance communication. For its part, what electricity displaced was the steam engine. With its capacity to be both here and there instantaneously, electric current made it possible to move the noisy, black-soot-belching behemoths we used to run much of our world (and still do) farther away from where we lived and worked. We were already making power and using it, yet with electricity we were slowly freed from the obligation to do both of these things in the same place. This was not only about moving power plants out of factories, but also mills away from waterwheels, streetcars away from horses, and light away from fire.

This capacity to translate power from one point to another took about seventeen years to become the cause célèbre of electricity. We can see these seventeen years—between 1879, when the first grid was built in San Francisco, and 1896, when the Niagara Falls power plant began sending its current twenty-two miles along high-voltage lines to Buffalo—as the history of a slow but steady technological comprehension. Over these almost two decades, the remarkable capacity of electricity to produce good light gave way to the more awe inspiring possibility of electric power. As each of these years drew to a close, America found herself home to ever more practical machines (and some kooky, bad ones) designed to render feasible remotely produced mechanical power and remotely fueled electric lighting. This was a rocky process, hardly linear, characterized by trial and error, a lot of guesswork, and experimentation without true models.

Because there were many stumbling blocks and puddlings along the way, the routes not taken are often overlooked in the story of how we got from the domestication of electricity in the 1830s, to its mass deployment in the 1880s and 1890s, and from there to the electricity dependent nation we now live in. These oft-overlooked developments are not, however, always secondary to the story.

In part this is because competition was so important to the ways in which early inventors plied their trades. Already by the 1870s most were trying to come up with viable products for obvious gaps in the market: a light that worked well for city streets was too terribly bright for office interiors; a circuit that worked well for small networks was disastrous for larger ones; a current that functioned perfectly well over short distances petered out when asked to travel further afield; a voltage that could light one hundred bulbs couldn't budge a streetcar. Every failing in the earliest marketable electricity systems also pointed the way toward a better path, a better solution, than the one that was foundering. This ebb and flow of inventive capacity is not, however, of interest only to the dedicated historian. Many of the earliest developments in electrical infrastructure—the systems we thought we'd left behind—are gaining unexpected purchase in the present as we remake and reimagine our grid.

For this reason it is worth charting a path to the grid we did get, first realized at Niagara but which would later grow to include the big federal dams (Bonneville (1937), Grand Coulee (1942), and the Tennessee Valley Authority project (1933)) as well as big, investor-owned, utility-run, government-regulated coal-burning, oil-burning, and later nuclear power plants. I will also, however, linger in consideration of the electrical power systems we once had but which did not for various reasons make the cut. Some of these, like arc lighting—a painfully bright form of early electrical illumination—would slowly fade from historical memory as it was outperformed by better, newer technologies. In arc lighting's case, the agents of its gradual demise were the parallel circuit and the much dimmer "soft, mellow, grateful to the eye" illumination produced by incandescent bulbs. The promise of technologies like the arc light was collapsed into obsolescence by the power of invention itself.

Other moments in the grid's history and other means of making light and power have had a longer afterlife than anyone might reasonably have expected. Individual power systems for a home, building, or factory were especially difficult to get rid of, surviving well into the 1920s, despite better technologies and strong utility and regulatory resistance to their continuation. These so-called "private plants" remained a part of our energy landscape mostly because people liked them and, almost as soon as the first wave of serious grid reform began in the late 1970s, they started popping back up—despite fifty years of total dormancy—to exert a newfound presence on the electricity scene.

Even more recently, with the meteoric rise of solar power installations, the preferred form of electrical current in the 1880s, so-called DC has also made a curious comeback despite having been outperformed by its most immediate competitor, AC, or alternating current, for 130 years. Although almost everything about our world has changed since the 1880s, it nevertheless remains the case that there is a certain appeal to private ownership, especially when explicitly linked to a sense of control over a limited domain. It just so happens that small, privately owned power systems are what DC is best suited to.

The early story of the grid, then, is one of rapid technological change. Inventing it was a collective process of learning what a newly domesticated electrical force was capable of and what things, small and large, might be made to realize its remarkable powers for the widest possible set of applications. This story is not so much one of winners and losers, but of moments of triumph and obsolescence as the speed of technological change outpaced even the most remarkably creative men who, upon stepping from the game, ceded their spots to others with new visions and newly necessary capacities.

We've been able to produce electricity using electromagnetic forces since Michael Faraday's experiments in the 1830s and to produce it chemically, using something like a battery, since Alessandro Volta invented the electrochemical pile in 1800. Despite an intense interest in electricity and the machines one might devise to make it, it remained unclear well into the 1860s what electricity might be good for. Such that by 1870 though we could produce electricity and control it, we had nothing to do with it. Many of the first generation of electrical product developers (inventors all, but with an eye toward serviceable, reproducible technologies for sale) suspected that the answer would be light. It was a dim, smoky era of itchy eyes, dark factories, and early bedtimes. Some sort of lighting better than the candles and gas lamps that dominated the era was on almost everyone's list of necessary improvements to business and daily life.

It was big news, then, when Father Joseph Neri, a professor at San Francisco's Saint Ignatius College, installed a small battery-powered electric light in his window in 1871; slightly less than a decade later, in 1879, San Francisco already had the nation's first-ever central arc lighting station. This system consisted of two dynamos (an early kind of electric generator) powered by a coal-fired steam engine at Fourth and Market Streets. It may have been tiny, lighting only twenty shockingly bright lamps, but it was a grid.

Less known, but no less remarkable, was the adoption of hydropowered electric lighting in the gold mines of the Sierra Nevada. The first electric grid built to light a mine also became operational in 1879. Using water-driven dynamos deigned by Charles Brush, the mine owners were able to light three 3,000-candlepower arc lamps. These allowed the miners to work through the night, effectively doubling the time they could spend searching for bright mountain gold. Electric lighting was so intimately linked to profit, in this case, that it quickly became an essential tool for mining companies. Slowly these companies also began to electrify pumps and hoists, building longer and longer lines to carry electric current from where the water fell to extraction sites. As electricity grew into an increasingly necessary form of industrial power, the grid's lines also grew and stretched to accommodate the circumstances into which it was built and within which it would need to function.

The press loved it too. Arguably no one needed electric light so much as clerks, writers, and seamstresses—people whose eyes were their livelihood—though the women hemmed and darned by the flicker of gas lamps long after electric lighting had become commonplace in the large office buildings that housed white-collar workers and newspapermen. The year 1879, which saw electricity come to the Sierra Nevada also saw an electric lighting system for the *San*

Francisco Chronicle: again, arc lamps, five of them, strung in a series. Three years later, in 1882, the *New York Times* had its offices wired for light, in this case fifty-two incandescent bulbs strung in parallel.

This subtle-seeming transition in the structure of circuitry, from serial to parallel, was the grid's first revolution. Though we tend to give Thomas Alva Edison the credit for having invented the lightbulb (he did not), he did devise something just as remarkable—the parallel circuit, one of his greatest if least lauded contributions to technological underpinnings of our modern world. The very existence of a relatively dim bulb, which we take for granted today, was made possible only by the prior invention of the parallel circuit.

The engineers, physicists, tinkerers, and inventors of Edison's era argued that the physical "division" of electric current was technologically impossible. That is, they had proved—before Edison found a workaround—that the notion of a bulb in every socket and a socket in every room was not only not financially feasible, but actually against the laws of nature. What was feasible was a power plant for every lighting system and a single or very limited number of fantastically bright lights for every power source. This is precisely how arc lighting works—each network consists of a short string of tremendously bright lights powered by a single generator. As natural as it might seem to us today that lots of bulbs, and lots of other stuff, would run off a single power source, this was not the most evident or even the most inevitable form for mass electrification to take in the late 1870s.

Edison, however, not being an educated man but rather a diligent one, had a tendency to disregard the laws of physics when pursuing technological solutions to irritating problems, occasionally with success and often with failure. Nikola Tesla, an inventor of a very different sort, once said of him, "If Edison had a needle to find in a haystack, he would proceed at once with the diligence of the bee to examine straw after straw until he found the object of his search . . . a little theory and a little calculation would have saved him 90 percent of his labor."

As it happens, Edison's diligence aside, the physicists were actually right. An arc lamp, because of its low resistance, can't be extensively divided. If a 1.5-horsepower dynamo ran the smallest of Charles Brush's arc lamps, then a system of eighteen such lamps strung in a series took all the power of his largest dynamo—a 13-horsepower monster of a machine. That is a lot of electricity and a lot of light. In fact, it was too much of both for people used to the sixteen foot-candles' worth of light (the equivalent of a 25-watt bulb) given off by the average indoor gas lamp. What is more, the color of arc lights was horrid, making everyone look "ghastly," and they buzzed disagreeably. The power frequency of arc lamps was, much like early fluorescent tubes, within the range of human hearing. That said, one could see quite well by arc lighting, and that was a decided improvement over the flickering flames of gas, candle, liquid coal, or whale oil lamps with which arc lighting was in direct competition.

In order to overcome this problem of too few, too bright, ghastly, buzzing lights, Edison adapted his parallel circuit, a rudimentary version, which had emerged from his work on the transmission of sound, to the problem of subdividing an electric current. At issue was that unlike water an electrical current doesn't seek the easiest or shortest route from one point to another; to electricity all pathways are equal. So if one provides two paths, it will take them both simultaneously and indiscriminately, even if the second is twenty times longer than the first; if one provides forty paths, this pattern of all-options-at-once travel is the same. The simplest explanation of the difference between the two kinds of circuits is that parallel circuits allow for this sort of both/and flow pattern, whereas series circuits give electricity only a single path to follow.

Older chains of Christmas tree lights work in this single-path way. If you have ever spent an evening trying to find the burned-out bulb in such a chain, you'll know that the downside of series circuits is that everything wired into the circuit needs to be able to allow the electric current to pass. One broken connection, one burned-out bulb, renders the entire circuit unusable. If we used series circuits in our houses today, all the lights and the fridge would shut off when you turned off your TV. With series circuits, either everything is on or everything is off.

Parallel circuits, in contrast, allow the electricity flowing through a system to take many possible paths. If all the paths have the same resistance, say a 15-watt bulb wired into them, the electricity in the system will take all the available routes simultaneously and without bias. If one of these bulbs burns out, the rest will continue to work as before; one path is blocked, but all other paths remain open.

If the maximum number of arc lights wired in series was eighteen, each burning with the equivalent brightness of a 2,000-watt bulb, all of which had to be either on or off, then Edison's grid, the grounding premise of which was the parallel circuit, looked a whole lot more like our own. When Thomas Edison put this theory into practice, building his first grid at the Pearl Street Station in Lower Manhattan, it was an

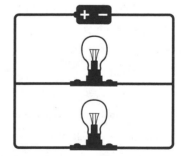

FIG 2 In a series circuit (*left*), if one bulb is out or off, the electric current cannot continue its path around the circuit. In this case, everything else wired into the circuit will, by necessity, also be off. In a parallel circuit (*right*), a burned-out bulb or other blockage does not make a difference to the movement of the current. Other devices can be turned on and off at will. (*Loïc Untereiner*)

act of competition. Charles Brush's arc light systems were all over the city and had been for almost three years. What Edison set out to prove at Pearl Street was not that electricity was good for making light—by 1882 that was known—but that electric lighting could be comfortably dim and electric systems greatly expanded.

When Edison's first public grid flickered to light in 1882 it was little more than a sixth of a square mile. By 1884, it had expanded to a full mile and held more than eight thousand bulbs, each a little circle of golden dimness with a lumen count equivalent to a contemporary 15-watt bulb. These eight thousand bulbs were wired in parallel, connected one to the next by a network of more than a hundred thousand feet of wire, which ran through conduits buried under New York's cobble. The entire network was powered by six 27-ton 100 kW Jumbo Mary-Anns, as Edison's dynamos were called for their admittedly "leggy" form. This was the nation's first public lighting system to rely exclusively on incandescent bulbs wired into a parallel circuit. It ran direct current and was coal fired—a messy way to make power in a place as densely populated as lower Manhattan, where horses laden with black dusty coal arrived in a constant train clogging the narrow street. While equestrian life was far from exceptional in the mess that was street life at the time, further increasing their numbers was a boon to neither the livability nor the traversability of the area around Pearl Street.

One power station was irritation enough, but Edison's plan for his grid was not limited to what he could accomplish at Pearl Street. Because of the limitations of direct current his network could not physically be extended farther than a mile without his having to build another power station. His vision for the city was a power station for every dense city mile, each with its clog of horse-drawn coal carts, its belched black smoke and its piped in source of fresh water.

Even though thousands of arc light systems were burning in American cities, manufactories, and mines before Edison turned the lights on for the staff of the *Times* and other residents of Wall Street, the Pearl Street Station is where, in the American popular imagination, the electric age began. This is due not so much to the fact that our current grid looks or works like Edison's first attempt—in many ways it does not—but because parallel circuits changed both the intensity of lighting and the proliferation of bulbs, and both of these have become ordinary to us in the present.

Though it's Edison's name that we remember today, rather than Mr. Charles Brush's—the inventor of the arc lamp and a dynamo that rivaled Edison's own—Brush's arc lights were, in the early 1880s, swiftly becoming the nation's preferred means of mechanical illumination. According to an 1881 issue of *Scientific American*, 1,240 arc lamps had been installed in woolen, cotton, linen, silk, and other factories, 800 in rolling mills, steel works, and similar machine works, 425 in large stores, hotels, and churches, 250 in parks, docks, and summer resorts, 275 in railroad depots and shops, 130 in mines and smelting works (like those in the Sierra Nevada), 380 lights in factories and "establishments of various kinds," and 1,500 along city streets, with another 1,200 lights in England and other foreign countries. Indeed, in England, where an Edison-style grid (lots of dim lights run on a parallel circuit) was very late to flourish, arc lighting was the preferred method of electric illumination well into the 1920s. Today, arc lights, with their inconvenient brightness, are used pretty much only as the source of illumination in IMAX film projectors.

Edison's system was not without competition, not only from arc lighting but also from gas lamps. Add to this that there were literally hundreds of dynamos on the market before Edison invented his own. His parallel circuit might have been revolutionary, his bulbs an aesthetic hit, but there was no real reason for his grid, his dynamo, or his recipe for insulating wires to be taken up by companies and municipalities considering electric lighting.

To give himself an edge, Edison took a route not unfamiliar to students of contemporary marketing. In creating his grid, he took care to make each dedicated piece only interoperable with other Edison components. If you wanted his bulb, or his circuit, then you had to also buy his dynamo, his switches, and his lines. When something went awry, you had to hire Edison's men to come fix it. Edison's grid was thus rather like a kit—a grid-kit—which could be customized for whomever wanted a full-on, ready-to-go, public or private lighting system. Many early customers were factories and large public buildings, some were municipalities, and very few, very occasionally were wealthy individuals.

The larger versions of Edison's parallel circuit system, installed by cities, were called "central station" grids because at the very heart of the network of lines stood the power station. The smaller, privately owned versions, which would plague latter attempts to unify electricity networks, were known as "private plants" because they had a single owner, and usually also, a single use. The owners of electrified mansions usually used the plant for lighting, while the managers of streetcar companies used them for the electrified hauling of persons all around the town.

In 1882, not long after Manhattan got its first incandescent grid at Pearl Street, a wealthy businessman from Appleton, Wisconsin decided to invest in a direct current Edison grid, which arrived by train with all necessary components, including the dynamo, and a team of men to do the installation. His grid-kit included a couple of generators, copper wires insulated in cotton, and incandescent light-bulbs with carbonized bamboo filaments. Though this was but one of a number of kits shipped out by Edison that year, what makes Appleton's case unique is that, unlike Detroit, New Orleans, St. Paul, Chicago, Philadelphia, and Brooklyn, all of which were early adopters of Edison's central station grid, it wasn't just industry, offices, or public spaces that were lit by Appleton's grid—it also lit houses. In

Appleton we got America's first true, small municipal grid, power for the people, and it ran not on coal but according to the erratic spinning of a waterwheel.

Like the Pearl Street Station, Appleton's grid produced and distributed direct current at 100 volts, give or take, and was used exclusively for powering incandescent lightbulbs—also of Edison's manufacture and included in the kit. What quickly became clear to the town's people and nascent utility men of Appleton is that maintaining a constant voltage is much easier with coal-powered dynamos, like the ones Edison was using in Manhattan, than with water-powered ones.

The problem with water in the Midwest and on the East Coast is that it is always falling at a different rate. After a spring deluge it is torrential; in a late summer drought it dribbles and drips its way downward. Curiously, this is less a problem in the West, where low-velocity, high-volume mountain streams run at a remarkably consistent rate as gradual snowpack melts meld into early-autumn rains. The Sierra Nevada was by its very nature a better place to install early hydropower, which ran in rivers and streams, than were Midwestern industry towns, because with every change in water's velocity, the intensity of the power produced by the dynamo in early hydroelectric systems shifted as well. As a result, the brightness of Appleton's bulbs was not entirely predictable. Too dim to see by at one moment and too bright by far the next, the vagaries of nature were maintaining an impolite quotient of control over man-made lighting.

In Appleton, this meant that everyone on the grid still needed candles and kerosene lamps on hand for when their bulbs faded to little more than a gentle orange curl of filament. For their part, the utility was constantly struggling to maintain sufficient funds for new bulbs, as the "too bright" produced by excessive voltage all too often burned out a bulb's delicate filament, rendering the entire contraption useless. At $1 a replacement bulb (or $23.50 in 2015 dollars), this voltage problem was not a negligible expense for tiny Appleton, Wisconsin's fledgling utility.

In order to understand the problem Appleton was having with its waterwheel and, more critically, the problem that all of America would be having with its grid by the mid-1880s, it is necessary to delve into the technicalities of electricity production, and voltage most especially. Though there are many ways to describe voltage, none make it a particularly easy concept to grasp. Often, again, what the process of trying to get down to the physical nuts and bolts of electric power makes most plain is how much electricity really isn't like anything else we know or manage.

So difficult is it to "think" about electricity that descriptive metaphors quickly enter into any explanatory matrix. Odd as it may seem, the most accurate of these for explaining voltage is the highly anthropomorphic notion of "desire": electrons that have been artificially split from atoms (which is what an electromagnetic generator does) "want" to resolve themselves back into whole atoms again. This may make electricity sound rather more like a singles bar than a problem of physics, but the notion of "desiring to complete oneself by coupling" does give a pretty good sense of the ferocity with which electrons without atoms move toward atoms that lack electrons.

What an electric grid does, then, is first forcibly divorce happy electrons which hold a negative charge from their atoms which hold a positive charge (generation) and then provide an easy route (the wires) for them to reunite again. As the electrons travel along these wires they pass through all the things we put in their way—things like incandescent lightbulbs. And as they pass they encounter resistance. A filament in a bulb is less conductive than the lines into and out of the device. Some of the electrons' potential—the push it has to reunite—is thus expended in getting through this resistive material.

In the case of an incandescent bulb this potential becomes heat, some of which radiates in the visible spectrum—about 5 percent for an Edison bulb. In an electric motor, like those in refrigerators or washing machines, this electrical stream is used to produce a rotation within a magnetic field, which in turn produces a mechanical force. This nicely boxed-up conversion of electrical energy to mechanical energy runs much of our modern world, from nose hair clippers to electric cars.

Regardless of whether one is siphoning off the "desire that moves them" in an electric flow for heat, light, or power (all of which are subsumed, technically, by the term "work"), two things are important: first, that the electrons are moving through, not stopping in, the device; and second, that they are not overly diminished in their desire (or potential for work) in the process. The name for the drive, also called a potentiality, is voltage. The measure of this drive is a volt—or a unit of electrical tension.

Voltage is measured by means of a potentiometer (it does, one must admit, sound like a poorly accomplished sex joke from start to finish). The strength of voltage gives us a good sense of the work that electrons can reliably be asked to do as they bump their way through the system and pass through our stuff.

One hundred ten volts can power a bulb, 220 an electric razor, 500 a streetcar, and 2,000 an electric chair; 50,000 volts is good for high-voltage transmission lines and 100,000 volts for a stun gun. Lightning, earth's wild electric force, is several hundred million volts, and best avoided.

As abstract as this all might sound, maintaining constant voltage on a grid is not merely a theoretical problem, nor even principally a financial one (Appleton's crippling bulb budget to the side). At root it is a practical one. If we want the grid to work *for* us, the relative vehemence of the electric flow must be controlled. For though we are accustomed to saying that our modern world relies upon electricity, it's far more accurate to say that it relies upon constant voltage, which electricity, when properly managed, provides. Our machines, unlike ourselves, brook no variation.

Or to put it another way, an electric grid's most significant point of appeal is its ability to make and reliably transmit standardized power—not for us, we are fine to get the candles out of the closet if the lights grow too dim. Our machines, the true customers of the electric age, are not so forgiving. Sagging voltage will tear a power plant to bits in minutes as all the tiny ball bearings that make that plant work quickly build up enough heat to destroy every motor and pulley system in the place. Less dramatically, a decrease in voltage will cause bulbs to dim, and the second hand on electric clocks to slow.

This laggardly voltage can happen for all sorts of reasons, though the most common in Edison's day was not enough "desire" generated at the plant to keep everything strung along the wires chugging along at a constant pace. The occasional sluggishness of Appleton's local waterwheel didn't cause electricity to cease to flow through the grid (a blackout), it simply didn't provide quite enough power to actually run everything wired into that grid at its full potential (a brownout)— a bulb that needs 110 volts to shine like "a little globe of sunshine" is, at 85 volts, a dim little beige sort of thing that turns all the world not from night to bright but into a sepia sort-of-colorless wash.

If tiny Appleton was struggling in 1882 with one kind of voltage problem, America's urban areas were fast confronting another. Different machines were being built to run on different voltages. This wasn't merely a product of inventors' whimsy but was linked rather to the amount of power it takes to make a machine work. In the 1800s an incandescent lightbulb, much like today, needed 100 to 110 volts to glow with adequate brightness, whereas a streetcar ran on 500 volts, and early electric factory motors ran at either 1,200 or 2,000 volts. This makes sense; it takes more work, and thus more energy, to move a wagon full of people along a rail, or to run a mechanical printing press, than it does to get a minuscule carbon filament to glow in a vacuum.

The problem for places like New York or Cleveland or Chicago (but not necessarily for Appleton) was that there were many potential uses for electricity, each with a specific voltage need. Direct current, which is what all grids ran until about 1886 and most grids ran well into the early years of the twentieth century, uses an inflexible voltage set at the dynamo. If a grid is using direct current, then a streetcar simply cannot be run using the same generator or set of wires as a string of lightbulbs. Either all the bulbs explode (overvoltage) or the streetcar doesn't budge (undervoltage); there is no happy medium.

The result was that in big cities, the grids almost immediately began to multiply. Direct current grids were not only built to meet different voltage needs of their customers, but competing electric companies and streetcar companies were unmotivated to share infra-structures, each preferring to construct a grid of its own. Each of these private grids had its own generating station and dedicated set of lines. Add the telephone and telegraph wires that were already occu-pying the poles these new grids shared (in much the same way that TV cable networks and telephone landlines share poles today with electric distribution networks), and the space above city streets was a mess. By 1890 one could hardly see the sky in downtown Manhattan, so thickly strung were the wires for each of these separate, parallel electrical systems loping between buildings and along streets.

To make matters worse, all these different electric systems were made up of components patented by a host of competing inventor-entrepreneurs. In the 1870s and 1880s, literally hundreds of patents were issued in the United States for dynamos of different builds and characteristics. And as downtowns were lit across the nation—Muncie, Fayetteville, Jackson, Topeka, Laramie, Chicago, Los Angeles, San Jose—their city governments, streetcar companies, and industrial concerns were all making quixotically individual

FIG 3 New York City during the Great Blizzard of 1888. The wires are a mix of telephone, telegraph, and electricity lines. (*Reproduced with permission from the Museum of the City of New York*)

choices about the best system to light their streets and power their machines. Some had arc lighting, others a version of the Thomas Edison's mail-order grid-in-a-boxcar that Appleton, Wisconsin, had installed, and by the end of the decade still others were putting in George Westinghouse's alternating current grids. By the 1890s, most had some combination of the three, plus large numbers of private plants as everyone from the brewer to the mayor had started buying and installing their own power systems.

The more electric power there was the more strongly a cultural proclivity rarely mentioned in histories of the grid was making itself known. Rather than opting to be hooked into a public grid, industrial concerns and some few wealthy people were choosing to make power for themselves, privately on site. From the 1880s well into the 1900s electricity production was actually tending toward becoming the sort of individualized system we use today to make hot water or cool air. The central station grid—as both a notion and a product—was receiving its greatest competition not from other electrical companies or competing electrical systems but from private plants—a grid no larger than a single home, or a single factory; a grid bought, paid for, and owned by one man, one family, or one industrial undertaking.

Though we tend to overlook it, hot water and cold air can, much like electricity, be produced more economically in densely populated areas by central stations than by individuals. In some places this system became the norm. The Soviet Union had municipal hot water plants that also made steam for heating, and many American universities have a central cooling facility that uses a campus wide system of ductworks to air-condition everything out of one central cold-air plant. There is no material or infrastructural reason that these two products are for the most part in the United States privately owned. The same was true of electricity. We think of it as a public good only because it became one with time, not because it worked that way from the beginning.

The history of the grid has been largely written to give the backstory for the central station, polyphase AC system we use today (and have been using pretty much exclusively since about 1915). But in the late 1800s during the scramble to light, and later power, as much of urban America as possible by means of electric current the story of our electrical infrastructure could have unfolded differently. Municipalities, investors, and inventors may have preferred the central station model, but people and factory owners liked small power systems. They liked a coal chute in the basement, a dynamo down there too, and a house or factory wired to use this locally produced power—mostly for light in the early days, but as electric motors became more reliable and electrical appliances more commonplace, these, too, could be powered on site. Nor did these undertakings need to exist inside a building. The grids that ran streetcars well into the 1920s were also essentially private plants—despite the sprawling geography these lines traversed, there was but a single customer and single producer for the electricity each used. Though we tend to think of electricity as an elite product during this epoch because only rich people could afford to have it in their homes, it might be more apt to say that it was a corporate and industrial power source first and a means of artificially illuminating the homes of the rich second.

Brush's arc lights, too bright for home use, sold as private plants at a rate of two to one over municipal sales. Similarly, Edison's central station grid-kit was selling best to single owners. One hundred and fifty of these were operating in the mansions of the rich like those of the Vanderbilts or Mr. J. Pierpont Morgan before Edison had even gotten the Pearl Street Station up and running. And once Pearl Street was working and the Edison General Electric Company had been established for the making and marketing of small municipal grids to cities across the Midwest, another 134 isolated plants had, much to Edison's chagrin, been ordered and installed, most into factories. "The light that proved most salable," one history recounts, was in the form Edison "considered unsound: the isolated plant proved to be as easy to sell as the central station was difficult. By the spring of 1883, there were 334 isolated plants in operation and the Pearl Street plant was still almost alone as a central station."

From then on, for every grid Edison franchised to a town he sold ten private plants. Between 1882 and 1887—the year that George Westinghouse's competing AC system entered the market, taking it by storm—Edison licensed 121 central station grids, like Appleton's. During the same period he'd also sold and helped install over 1,200 private plants. And, to jump ahead a bit, in 1907 only about 8 percent of American homes were served by electricity and most of these were still private plants—this, long after the chaos in the skies had been resolved, by means of the rotary converter, into a "universal" system for electricity production and distribution.

Part of the reason for this is that as remarkable as Edison's direct current grid was in densely populated lower Manhattan or tiny Appleton, Wisconsin, it was better suited to small installations than large ones. The Pearl Street Station only confirmed this, as even with the six jumbo dynamos he'd installed it still didn't have the capacity to transmit electricity farther than about a mile. He could put a lot of bulbs in that mile, and by adding additional generation, he could power them well, but after that the laws of physics took over. Direct current at so low a voltage (100 volts) just can't be transmitted farther before becoming so diminished that one couldn't singe a mustache hair with the stuff. There just isn't enough *oomph* in the electron stream to push it beyond a mile.

If limiting for mass-lighting or large municipal projects, this characteristic of direct current meant that it was perfectly suited to smaller scale installations. Even a big factory, or an office tower, or a rich man's mansion were, in the late 1800s, smaller than a square mile. It was a manageable task to determine the light or power needs of each individual customer and build that customer a basement power plant, install the necessary lighting, drill holes in the walls for the electric sconces, add the bulbs and switches, and call it a success.

Strictly speaking there was no reason to engineer bigger, more complex grids than this. Lots of small grids were the order of the day. One suspects, though it is conjecture, that had these private plants become the way we made and also thought about electricity, the chaotic tangles of wires that bedecked nineteenth-century city skies would have calmed in time as traction companies consolidated, as competition between local power providers eased, and as municipal grids gave way to smaller building-based electrical installations.

This didn't happen, but an understanding of a past that almost, but didn't produce an electric system based in private plants rather than central stations helps to illuminate some of the peculiarities of the present.

Today, the tension between individual power production (private plants) and utility-supplied electricity (central stations) is once again becoming the battleground upon which the future of our grid—the form and scope of its infrastructure—is being waged. For most of the intervening years, which is to say, the entirety of the twentieth century, it seemed natural that power would be made en masse, transmitted by immense, high-voltage AC lines (and, by the late 1960s, high-voltage DC lines like the Western Doughnut as well), to clumps of customers, linked one to another and to the greater grid through a massive system of pole-top transformers and local substations where the intensity of transmission scale voltage was stepped down to domestically consumable flows.

Until, by the turning of the most recent millennium, this sense of a big public grid as the natural way to make and distribute electricity was, for all intents and purposes, absolute. Even when I spoke to people who were so upset with the inconsistent functioning of utility-provided electricity that they were trying to figure out ways to make their own power for themselves, they nevertheless consistently reiterated the view that electricity was a basic human right. It was something the government should ensure all people had access to just like potable water or breathable air. Indeed, this is also my view.

The most diabolical outcome of a return to a system of private plants, which could easily happen in the next couple of decades in sunny places like Arizona, Hawaii, and southern California (and to some degree has already happened in Germany) is that it threatens universal access to quality electrical power. If our nation is grounded in the notion of equal opportunity for all, then there cannot be partial, or unevenly distributed, electricity. We cannot continue to be the nation we have become and also endorse or produce haves and have-nots in the world electric. One can go even further to claim that it is not only electricity-for-all which makes our country strong, nor even electricity priced so low that almost everyone can use it with unbridled avarice, but that we all have access to the same *quality* of affordable electric power. The poor aren't faced with flickering bulbs, or a ration system that grants them an hour of electric power per day. They are not saddled with prepaid meters, like those in South Africa, that function only if one has made a deposit on the account before use, while the rich have air-conditioning, light, Internet, and electronic trading systems 24/7. Standardized power for all is a twentieth-century value that has done much to make America what it is. As private plants come back into vogue, it will be important to keep in mind how shifts in power production from a big grid into many small ones might be integrated into a system that still promises and still provides high-quality electricity to everyone.

In the early days of electrification, however, this sense of power as a common good was not how electricity was made or marketed. It was by definition an elite product, not for everybody but for those who could afford it. This was not simply because electricity hadn't yet spread to the masses, but because the masses were not considered a market worth reaching until later in our nation's history. Some even argue that it wasn't until after the Great Depression that the notion that one could make money by selling lots of cheap things to lots of relatively poor people made any inroads at all. Most, however, agree that the idea of a "mass market" or "consumer culture" came about in part because utility company entrepreneurs needed new ways to make a profit from electricity. GE (originally Edison General Electric) started out as a power company; it was only with time that it became an appliance company renting, and then later selling, people stuff that needed to be plugged in to work. Money was made twice over, first with the sale of the refrigerator and second with the ongoing sale of electric power necessary to make it run. This shift in vision involved both imagining a whole new set of demands, beyond lighting for elites, for which electricity might be used and seeing a whole new population—everyone—as a potential market. Not elite light but popular light, popular heat, and popular power.

In other words, the notion of consumer or mass culture, in which all people are promised access to all things, was in part the result of the universalization of electricity and not the other way around. A standard-ized electric current came first, with the construction of the Niagara generating station at Niagara Falls in the early 1890s, while the veri-table explosion of home appliances, and the idea of a mass market for them, did not emerge until the early 1940s. The standardized plug and outlet that made this consumer revolution possible was not achieved until the unfortunate year of 1929, after which point—the Great Depression decimating our lives and livelihoods—very few people in America were in the market for domestic technologies of any kind.

The first step toward a big grid, one that would make it possible to universalize access to electric power, was the invention and successful manufacture of alternating current (AC) electrical systems in 1887. It seemed at first that adding yet another different sort of electric current to the chaos multiplying at rooftop and pole top would only make things worse, and for a while that is precisely what it did. But AC had a strong advantage over DC, an advantage that with time facilitated the streamlining and ordering of the tangle of wires, companies, currents, voltages, and rates of oscillation. For unlike direct current, alternating current can travel; it was outpacing its rival's single sad mile from its first moment on the market.

The singular advantage of alternating current is that low voltages, made at the generator, can be "stepped up" to much higher voltages by means of a transformer—a simple device made of two sets of tightly coiled copper wires that almost, but don't quite, touch. Higher voltages can go farther than lower voltages. It has, if you will, a higher quotient of desire, it "wants" harder, and thus is propelled farther. The transformer is a simple if genius means by which this "stepping up" and "stepping down" of voltage is accomplished without any loss in efficiency. It only works, however, if the current traveling through it "alternates," which is to say, moves in a series of rapidly reversing waves.

Despite the fact that their flows move somewhat differently the underlying physics of alternating current and direct current are the same. In both, electrons are forcibly separated from atoms, and once the divorce is accomplished, the former seeks out the latter with a magnetic sort of ardor. The difference is that with direct current, the direction in which this positively charged slot (in an atom) is to be found is always the same, and thus the stream of electrons seeking this slot always moves in that direction. In this case, electricity in a wire is rather like water flowing in a river.

With alternating current, in contrast, a rotating polarity at the generator causes the "direction" of the positive charge to switch. The free electrons, still passionate in their struggle to regain atoms of their own, thus also shift: surging first forward, then stopping, then surging backward. And stop. And forward. Stop. Backward. Stop. And so on and on and on. This flow of electrons is still doing work every time it passes through a machine. With AC, however, sometimes the electrons are moving in one direction through the machine and sometimes in the other direction. These stop/go, stop/go, stop/go reversals and interruptions in the direction and flow of power are so quick (in the United States, about sixty per second), that nothing much of what is plugged in even notices these directional shifts, though the buzzing of early fluorescent lights was a symptom of it. In the early days of the grid there were several types of alternating current—single phase, dual phase, and polyphase; the differences between these, while important, can be simplified for our purposes to the fact that dual phase and polyphase send multiple waves though the wires and single phase sends just one.

If Edison's Manhattan grid, in 1884, made and transmitted direct current at 110 volts over about a mile before its usefulness diminished to zero, Westinghouse and Tesla's early alternating current system, in 1886, made electricity at 500 volts but transmitted it at 3,000 volts. By 1891, AC had been transmitted 16 miles between Tivoli and Rome, 14 miles in Portland, Oregon, 2.6 miles in Telluride, and an extraordinary 108 miles at 40,000 volts in Germany. Every experimental installation from the late 1880s to the early 1890s proved more and more successful on this account, until by 1894, 80 percent of the grids being ordered and installed in the United States were using alternating current.

Just as arc lighting was not immediately replaced when lower voltage grids became the norm, direct current networks lingered on as alternating current systems grew in popularity. Each simply added to the chaos of systems, voltages, dynamos, alternators, lines, bulbs, and the thousand and one other components that multiplied in the wake of electricity's growing popularity. Among the proliferating elements, not even one set of common pairings stood out from the rest. In other words, there was no grid, but only hosts of individually, corporately, or municipally owned grids, all overlapping in the sky as on the ground.

Downtown Manhattan alone would have 1,500 arc lights by 1893, as well as twenty light and telegraph companies, each running independent sets of wires carrying varying voltages (if direct current systems) or operating at distinct rates of oscillation (if alternating current systems). Chicago, at roughly the same time, was home to forty different electric companies, most of which were associated with competing streetcar lines. Direct current operations offered power at 100, 110, 220, 500, 600, 1,200, and 2,000 volts, with one or more dedicated systems of wires for each. Alternating current, all the rage by the early 1890s, was not the immediate improvement it had at first seemed. It could be transported much farther and be made to change voltages fairly easily, a clear improvement over DC power, but in the early days there was little agreement about the optimal frequency at which this current would "alternate." As a result even this more flexible voltage system had dedicated wires for 25, 30, $33\frac{1}{3}$, 40, 50, 60, $66\frac{2}{3}$, $83\frac{1}{3}$, and 125 cycles per second. Making it all even worse, when electric companies went out of business, they simply left their wires to molder and fray in the sky.

By the early 1890s regular people, industrialists, white-collar workers, small business owners, corporate managers, manufacturers, traction companies, investors, and inventors had all grown fond of electricity and what it could do. Nevertheless, most also found the wild way of building an infrastructure deeply unsatisfactory, with four different kinds of current (DC, single-phase AC, dual-phase AC, and polyphase AC), two different intensities of lighting system (arc and incandescent) that relied on different wiring logics (series and parallel), seven different possible voltages, nine different rates of oscillation, and up to forty competing electric companies (depending on the size of the market), plus all the folks who'd opted out of public power and installed private plants. The result was predictable: not only were there too many wires, but the lack of standardization made it difficult for anyone to make any money, whether selling current or selling the various components and machines that used it.

There was no hope of cornering a market and it was equally difficult to manufacture machines that could work within this chaos. In the late 1800s America was an intensely industrializing nation. Automation and complex machinery were critical to this process and electricity seemed to promise a common currency by which these mechanical systems might be run. This promise was frustrated, however, by the intense, and increasingly irritating, diversity at the level of infrastructure and the lack of translatability between the various systems. This dissatisfaction, without necessarily prompting much agreement or kindly feelings between competitors, produced a willing market for one kind of machine in particular: the rotary converter—"a single armature for changing direct current first into polyphase and then the reverse."

Like the parallel circuit, this little thing has remained mostly unlauded in the face of more luminous undertakings. And like the parallel circuit it was a missing link without which the evolution of electrical infrastructure was not progressing very well. America was in this brief moment before the arrival of the rotary converter in danger of having nothing like a national grid, nothing like municipal grids, but just a mess of competing interests and inventions, mechanical systems, and investor preferences.

In 1891, when the decision was made to build America's first large-scale power plant at Niagara Falls and to send its power to Buffalo, some twenty miles distant, it was not even clear to the system's designers if that power would take the form of electricity. Even before coming to the question of which kind of current they would use, the Cataract Construction Company at Niagara had to first rule out tried-and-true mechanical means for transmitting power—by water under pressure, by compressed air, or by manila rope. They even considered not transporting the power produced at the falls at all, but rather using all 200,000 horsepower in the tiny village of Niagara Falls, population five thousand in 1890. How, they wondered, might they turn this hamlet into a manufacturing center for the Northeast?

Once it had been decided, toward the end of 1891, that moving power to Buffalo by means of electricity was the most efficient option available, the project's engineers did not immediately rule out direct current. It was not yet known, for example, whether several AC generators could be made to run in tandem with one another—the oscillation rate of each melding with that of all the others. This capacity, which Edison had demonstrated in New York, worked well with DC, as it allowed for more, or less, power to be brought online as needed. One could just pack a power plant with a number of different-sized dynamos and by turning them on and off in different configurations ensure that demand would always be balanced with supply. How, with AC, would similar changes in demand be met? The designers of the Niagara plant had no idea. Nor did they know if all the power they were hoping to generate at Niagara would be used for lighting or if other kinds of machines and motors would eventually draw from it. Indeed, so little confidence was accorded the capacities of AC power at this scale that Lord Kelvin, a physicist of some note, sent a cable late in 1893, just as the International Niagara Commission was completing its decision-making process, saying TRUST YOU AVOID GIGANTIC MISTAKE OF ADOPTION OF ALTERNATING CURRENT.

Less than a month later, two-phase AC oscillating at 25 cycles per second was decided upon and the construction of the Niagara hydroelectric plant, its generators, and its lines was begun. By 1895 the plant itself was complete and was supplying power locally, to tiny Niagara Falls; in November 1896, point-to-point transmission began to Buffalo, where the availability of constant power almost immediately began attracting large manufactories making things like abrasives, silicon, and graphite on an industrial scale. Perhaps most important to America's future, it was in Buffalo, with the force of Niagara's own power, where the manufacture of aluminum first became a profitable enterprise. Finally we had a light, strong metal for our airborne and land-bound engines; the era of the automobile was officially begun with air travel soon to follow.

The grid we would get, the grid that the mess of the 1880s would resolve into, reflected in many ways the decisions made at

Niagra—polyphase alternating current, oscillating at 60 cycles a second, produced by large power stations, transmitted by means of point-to-point high-voltage wires, and distributed by networked and ringed lower-voltage delivery systems, standardized at 110 and 220 volts. As more new electrical installations were built to follow this model, rotary convertors, and slightly later, phase converters (which made single-phase AC systems interoperable with multiphase systems) made diverse, existing electrical infrastructures interoperable, such that companies did not have to abandon large, unamortized investments in order to work together.

Traction (streetcar) companies got to keep their direct current systems while also making and selling power to neighboring AC grids; manufacturers got to keep making single-phase AC motors that could now work on polyphase AC systems; and polyphase AC could be made to work with smaller-scale, local DC distribution networks. Even the very first, smallest grids, those that powered arc lights, could be integrated into this mishmash of newly translatable technological systems. The old could be coupled with the new, and the mess of America's electrical infrastructure began to resolve itself, not in the direction of private plants, but toward what would come to be known as the "universal system" of big interlocking grids powered by central stations.

The hydroelectric plant at Niagara Falls was thus the closing bell on the effervescent, chaotic, immensely creative and inventive activity of the previous seventeen years: 1879, the first arc light grid in San Francisco; 1882, the first low-voltage direct current grid in New York; 1887, the first alternating current grid; 1891, proven long-distance high-voltage transmission. And in 1896, the completion of the first large-scale generating station at Niagara Falls, together with the first long-distance transmission wires in constant use, the total adoption of parallel circuits, incandescent lighting, and the equally near total adoption of alternating current. America had her grid.

This story of electricity's domestication and quick spread makes it sound as if it were common by the dawning of the last century for a home, a business, or a factory to have grid-provided electricity. It was not. Rural people had virtually no access to electricity until the passage of the Rural Electrification Act in 1936 during the darkest days of the Great Depression. And even before that, during the first decade of the twentieth century, urbanites, suburbanites, and factory owners rarely chose to use electric current over more constant and familiar technologies such as gas lamps and steam engines.

In 1900 only one factory in thirteen used electric motors, and only one domestic light in twenty was electrified—the rest were still gas lamps, kerosene lamps, and candles. At home, people often preferred the wink and flicker of fire to the more even, unwavering, if equally warm, light of incandescent bulbs. In factories there was some suspicion about the reliability of a steam-generated electric current when steam-driven mechanical engines had for so long worked so well. Add to this the fact that there were no outlets, no plugs, and very few uses for electricity other than for light—at home and in the office—and power in industry. In 1901 there were only eighteen refrigerators in all of Manhattan, and a decade later, in 1910, though there were close to 45,000 electrical appliances in Southern California, 80 percent of these were irons, all of which screwed into light sockets. Nor was the dearth of electric lamps, electrified motors, electric refrigeration, electric outlets, and the plugs that fit into them only about personal preference or technological lag. The state of state-of-the-art technology and the adoption thereof also reflected American social structures, income distribution, and business imaginaries at that time—nobody had yet grasped that electricity might be a product for the masses or that a utility could make as much money manufacturing electrical appliances as making and distributing an electric current. As much as we take universal access to electricity for granted, a mere century ago we were still a long way from mass electrification, in our minds every bit as much as in our infrastructure.

NOTABLE PASSINGS

The Pearl Street Station burned down in 1890; its technology, so recently state of the art, was hardly competitive upon its demise, and so it was not rebuilt. The site is currently a parking lot run by an affable native of New Delhi, India. Appleton, Wisconsin's utility went bankrupt in 1896 and was replaced by the Wisconsin Energy Corporation (now called "We Energy"). In 1956 the Niagara Falls generating station fell off the side of the cliff at the falls' edge, where it had been perched. A total loss, it was one of the largest industrial accidents in America to that date. John Haney, a janitor at the plant that day, remembers that "there was water seeping in and we were trying to keep it away from the generators. The pressure on the building was tremendous. The windows facing the river were just popping out, the concrete floor would buckle up and I would jump over it. Then the wall toward the falls came crashing down. I then headed to the elevator. Water and stone was falling into the bay. I stood under the steel doorframe watching the operator call to me and I jumped on. It took forty-five seconds to reach the top and I saw the gorge collapse on the area I was at." Thirty-nine men escaped the plant as it fell, while one was "hurled through the window into the river" when a "jet like a burst of water from a broken penstock carried his body along with thousands of tons of debris into the surging maelstrom." Power at Niagara is now made by the Robert Moses Niagara Power Plant.

CHAPTER 3

The Consolidation of Power

That first remarkable moment when the Pearl Street Station breached the interminable dusk of interior office space with the warm glow of electric lighting was not the only big news in 1882. This was also the year that saw the creation of the Standard Oil Trust, a business entity that brought 90 percent of the world's oil production and refining under the control of a nine-person board, headed by John D. Rockefeller. With twenty thousand oil wells, most in Pennsylvania, four thousand miles of pipeline, five thousand tank cars, and over one hundred thousand employees, Standard Oil was a thorough monopoly. Rockefeller's sprawling vertically integrated corporation perfectly captured the trend in the late 1800s of replacing many small companies in competition with one another with single large entities capable of producing and moving products to market at a scale inconceivable to family businesses or local undertakings.

The electricity business might have been a chaotic diversity of competing interests and alternate systems in the final decades of the nineteenth century, but it was the exception not the rule. According to the historian Richard Hirsh, in the years immediately following the creation of the Standard Oil Trust, more than 4,000 businesses in the United States were combined into 257 corporations, and "by 1904, one percent of American companies controlled 45 percent of manufactured products." These included corporations of phenomenal reach such as U.S. Steel, American Tobacco, DuPont, Anaconda Copper, and American Telephone & Telegraph (AT&T).

Like all big business undertakings these companies had the capacity to streamline their industries by binding more efficient production and transportation to economies of scale. As monopolies they also had the capacity to set prices since no customer could purchase oil, steel, copper, or telephone service anywhere else. Until it was dismantled by the U.S. government in 1911, Standard Oil was virtually the only option for almost any petroleum product then in existence—from kerosene, to fuel oil, to Vaseline. Mr. Rockefeller, in the process, became the richest man in the world. J. P. Morgan, the owner of U.S. Steel, became one of America's most august tycoons ruling over empires—including the one built by Thomas Edison—far beyond the Steel Industry. And J. B. Duke, the owner of American Tobacco, if somewhat farther down the list, still numbers the 49th-richest man in the history of the nation (for context, Warren Buffett is number 39).

Despite the reach of these titans of capitalism, monopolies didn't need to be national in scope or even big. Smaller, more local enterprises, like the grain elevator owners in the Midwest accused of price fixing in 1877, or even a ferry company with a single boat—if this was the sole means to transport goods and persons across a body of water—were effectively monopolies, in that they maintained absolute control of a market.

The most important thing about a monopoly or other organization of companies that colludes to fix prices is that the price of a good or service need not be related to the cost of producing it. A farmer with grain to sell also needs to transport and store it. He either pays what the elevator cartel is asking or his grain rots and he loses everything. Or, if he is slightly better off, he can build a silo of his own. For that, he needs steel; he pays the price Mr. Morgan is asking, or he doesn't use that metal in his construction. A customer's choice is limited to paying the price on offer, regardless of how arbitrary or exorbitant, or not partaking of that particular good or service and thus courting his own ruin.

Originally, as Hirsh points out, there was little worry that electricity providers might join in this trend. The early technology didn't permit it. Before the widespread adoption of alternating current, the physics of electricity production, and most especially distribution, limited both the scope and the avarice of any given electric company. Low-voltage direct current was by its nature a product that facilitated local owner- ship and the formation of separate companies for every use to which it might be put. One company might provide light, another factory power; a third would run streetcars, while a handful of others designed and built the dynamos, switches, and lines the first three depended upon.

Despite Edison's dream of creating an electric grid that was both vertically and horizontally integrated, his DC network was the exact opposite—decentralized, small, unwieldy, and prohibitively expensive to scale up. As such, electric companies provided very poor fodder for the processes of monopolization so popular in other industries at the time. Even after the advent of alternating current in the late 1880s, which introduced flexible voltage and vastly increased the scope of a given network, multiple, overlapping, and competing service providers were the norm.

As the nineteenth century rolled over into the twentieth and the universal system that had been developed in Niagara slowly became the standard for any new electrical project, there were still tens of companies making electricity in any large municipality. While small cities and towns tended toward municipal grids, in most cases these were run by local governments, nascent utility entrepreneurs, or mill owners who sold their excess power to nearby customers as a side product. And, just as most American towns in 1900 existed as thickets of population surrounded on all sides by miles of rural farmland, forest, or uncultivated prairie, so, too, were their grids tiny islands of networked power in a vast sea of flickering candle and gas lamps. In 1902 there were 815 city-owned and -governed municipal power companies in the United States, which grew at the rate of about 100 a year until 1907, when these numbered over a thousand and accounted for about 30 percent of the nation's electrical supply. Almost all of the rest of the electricity generated in the United States was made by private plants—mostly traction companies to run their streetcars and electric trains, industrial manufactories, and commercial buildings.

As late as 1905, it was absurd to imagine that one day the kind of electrical services that small-town America provided to its relatively isolated pockets of population might be knitted into a single system by monopolizing electricity trusts. Or that all the many streetcar companies functioning in America's bigger cities might be convinced to start buying their power rather than making it for themselves. And yet within twenty years the eight largest utility holding companies in the United States controlled three quarters of the electricity market. So powerful had these trusts become, and so questionable were some of their basic accounting practices, that not only did several collapse completely during the 1929 stock market crash, but the last wave of antitrust laws passed in the early years of the Great Depression was aimed squarely at them. The thrust of the 1935 Public Utility Holding Company Act was, as its title makes clear, to make it illegal for any company to hide their debts in "holding companies"—special shell corporations that made the actual worth of an enterprise impossible for investors to determine.

By 1925 almost nobody in the electricity business could even imagine a system for making, transmitting, distributing, or managing electric power other than as a monopoly enterprise. This was an extraordinarily rapid transition from chaos and competition to a single service provider. Remarkably disparate interests, including advocates of municipal power networks, of public power projects, and even electricity cooperatives, were all convinced by the 1920s that the monopoly was the best way to manage the manufacture and sale of electric power. There were different visions of who might control such a monopoly—the people, the government, a corporation, a co-op— but the basic notion that electricity markets were to be consolidated and managed without competition was, by the 1920s, universal.

Much of the credit for this change goes to the hard, thoughtful work (as well as some barely legal machinations with piles of cash behind the scenes) of Samuel Insull. Born and raised in England, Insull spent his first twenty years in America as Thomas Edison's personal secretary. Apparently his success in this job was attributed not only to his fastidiousness in all things, but also to the fact that he was one of the few people who needed as little sleep as the inventor. If Edison wanted to order a bunch of copper at two in the morning, Insull was at his desk, ready to make it happen.

As electricity shifted from being a game inventors could win to one better left to the balance sheets and business strategies of businessmen, Edison retreated from the limelight, leaving Insull to manage more and more of the inventor's holdings. By 1892, when Edison had to admit defeat in the battle between his DC system and Westinghouse and Tesla's AC system, retiring to his New Jersey laboratory to work on fluorescent lighting (a project that ended badly for him, blinding him in one eye and killing an assistant), Insull stepped firmly out from his august patron's shadow and took up the standard of American power.

Neither an inventor nor an ideologue—both terms that might be aptly applied to Mr. Edison—Samuel Insull was an accomplished businessman, and the business of electricity in the early twentieth century was ripe for innovation and, more important, for monopolization. There was money to be made, but only if the fractious nature of the industry could be resolved into a far less competitive, far more secure system, free of redundant investments in overlapping infrastructure. Insull wanted the kind of monopoly that Rockefeller had made with Standard Oil and J. P. Morgan had made with U.S. Steel. The best he'd manage was *almost* this kind of monopoly because of the ways in which his product was crucially not the same as theirs. Molding the electricity business into the form du jour of American monopolists was one of the most remarkable projects, intellectually as well as fiscally, of the twentieth century.

Insull didn't start out knowing how to do this. He felt his way forward with his gut, his intellect, and a very good sense of the relationship, in the late 1890s, between government, banks, and the business community. His first major stumbling block was the intractability of electricity itself: since it can't be stored it can't be stockpiled; since it is indivisible, it is difficult to count and accurately bill for; since it is lethal, it requires a highly trained workforce to manage; and since it is utterly inseparable from the infrastructure that carries it, one has to bear the cost of building and maintaining its infrastructure.

Given all of this, electricity was at the very least an extremely unwieldy commodity; at worst it was an almost impossible one to stay in business supplying. This was one reason why many early electric companies were so eager to sell and install private plants: it was an easier route to sure income than convincing a whole neighborhood to buy their product and then building the necessary infrastructure and maintaining the necessary workforce to supply central station power.

In the earliest days of mass-electrification the intractabilities that caused electricity not to work like oil, gas, steel, or grain seemed like minor issues, but they were not. Edison himself, as the historian Maury Klein explains, "discovered his error in using the gas industry as a model. Gas could be stored, which made it possible to produce on an orderly, rational basis like other manufactured products. It could maintain reserves to meet peak requirements and level out demand over a twenty-four-hour period. Not so electricity. It had to be produced, sold, delivered, and used all at once, which meant that the plant supplying it needed the capacity to deliver the total maximum load demanded by customers at any given moment."

In other words, one could build an electric monopoly and totally control a market, but first one needed to understand how the various pieces of that market fit together. An electricity monopoly does not have the same shape as a monopoly on grain storage or steel production. Because electricity is a temporal product, customers' consumption also had to be arranged, in roughly equal measure, around the clock. If profitability is the goal, midnight load needs to be on par with five P.M. load and ten A.M. load. Nothing else on the market demands this kind of leveling of consumption for each hour of the day, neither in Insull's time nor in our own.

Thus, though Insull proceeded like other monopolists in his dealings with competitors—mostly by buying them, or running them out of business, or buying the companies that supplied them (and thus running them out of business)—the secret to making a fortune off of electricity was not simply to have lots of customers but to have lots of different kinds of customers in order to provide sufficient demand to run a large, centralized generating station 24/7.

When Edison retired to his inventor's studio, Insull was offered a handsomely paid position as second vice president of the newly formed General Electric in New York—a position he refused. Insull was done being second, even if for one of the largest companies in the nation. Instead, at thirty-two years old, he packed up and moved to Chicago where, for $12,000 a year (a third of what GE had offered him), he took the helm of that city's Edison franchise. He was, at long last, the boss, albeit of a minority player, in a crowded market, in a foreign city that he made no secret of detesting. "His most vivid image of the city concerned neither its men nor its future," one biographer recounts, "but its filth and its huge rats." So worried was Insull that he would be driven by his distaste for Chicago to escape east-ward, back to New York or even so far as his home in London, that he insisted on a binding three-year contract. For that length of time at least, Chicago Edison was his and he would grow it on his own terms. He stayed much longer and did much more with this tiny franchise than even he could have imagined at the start, staying in Chicago until 1932, when he fled the country on government corruption charges. In the intervening years he would become the world's first utility tycoon, Chicago Edison his empire.

Upon Insull's arrival in the City of Broad Shoulders in 1892 there was hardly a more apt word for Chicago Edison than minor. Though they had opened the city's first central generating station in 1888, a 3,200-kilowatt DC power plant on West Adams Street, by the time of Insull's arrival on the scene there were eighteen central station electricity providers in Chicago's downtown Loop alone plus another five hundred private plants.

What this meant, practically speaking, was that Chicago Edison's five thousand or so customers only used, and paid for, all the power the Adams Street plant made during the early evenings. When dusk settled over the city, every front-office clerk and every corner-office executive alike found themselves in need of artificial illumination. The demand for electricity then dropped off precipitously as offices closed up for the night and the last of the city's workers stepped aboard L trains bound for the suburbs where they read by gas light and ate food cooked with a gas flame. Even the most precocious elec-trical entrepreneurs in the 1890s were limited in the scope of their dreaming to Chicago's urban center by the physical limitations of DC networks. The suburbs, and even closer residential neighborhoods, were too far away for electric lights.

At night, in the Loop, when no one was at work, in the mornings, and for most of the day (most especially during the summer), Chicago Edison's sole power plant, fully capable of supplying its 3,200 kilowatts all the time, sat idle or was massively underutilized. As Insull once famously said: "If your entire plant is only in use 5.5 percent of the time, it is only a question of when you will be in the hands of a receiver." He needed a way to sell power the rest of the day or his company would founder.

Before Insull's arrival, Chicago Edison, like almost all of Edison's utility franchises, had made up for the veritable impossibility of ever making a profit with a central station grid by selling, installing, and providing maintenance for private plants. Indeed a great many of the private plants in the Loop were variously sized versions of Edison's original grid-kit, which had been sold and installed by the franchise. While this was a wise strategy for maintaining solvency over the short term, over the long term the multiplication of private plants further diminished the customer base for central station power. These private plants, however, were in a critical way as limited as the Adams Street Station, then Chicago Edison's largest power producer.

Chicago's manufactories, for example, produced and used power during the day, turning their generators off at night; private plants in apartment buildings and luxury residences usually sprang into use in the evenings just as the bulb system in the business district was being shut down; streetlights—often municipally owned—only burned at night; while streetcars ran most intensively at dawn and dusk. Everybody was using their fairly expensive, almost identical infrastructure only part of the time because it was as cheap to produce their own power as it was to buy it from any centralized source.

In the 1890s and early 1900s, the sound logic argued that because of the relationship between capital costs, which were prodigious, and profit per customer, which was minimal, electricity was by its very nature an elite product. The only real way to make money selling it was to sell a lot of it at a high price to relatively few customers. It didn't matter to most central station managers that electricity cost 50 percent more than gas in urban markets, because electric companies were not in competition with gas. Gas was for the masses; electricity for the few. If in 1892 Chicago had a million people, fewer than 5,000 of them used electricity at home. Some optimistic souls thought that one day Chicago might have as many as 25,000 regular users of electric current. Such expansion was unlikely, but not impossible.

A short fourteen years later Insull's Edison had doubled the Pollyanna's estimate with 50,000 paying customers; by 1913 he had an improbable 200,000 customers, or a tenth of the population of the city.

What Insull wanted and strove to build was an infrastructure the inverse of what he was saddled with upon his arrival in the Midwest. Instead of many little generating stations, with many owners, running intermittently, he wanted one that he owned and which ran all the time. In order to do this he needed to acquire "load" for each time period during the day. He needed streetcar companies to buy from him at dusk and dawn, residential customers for the late evenings and early nights, municipal street lights for nighttime, businesses for the late afternoons and early evenings, and most important of all, industry for midday. He wanted to make a lot more power, make it round the clock, and to sell it all—every last watt.

The factories were the one piece of the puzzle that could make this scheme work. They were the only customers which could be counted on to use a tremendous amount of power during daylight hours; without factories, no vision of a single on-all-the-time grid could be profitable.

It's worth noting that even this ideal cocktail of customers and times of day fails to provide significant nighttime load. This remains a problem for our utilities today. Even taking into account public street lighting, electricity use drops off precipitously as people start to go to bed and only starts to creep upward again around six A.M. One of the reasons that electric cars have received such public praise is that they can be programmed to charge almost exclusively at night and thus provide that rarest of beasts—substantial midnight load. Insull, of course, gets no credit for the cars, but many of the other things we use electricity for at night we can thank him for zealously promoting, including a rate structure that rewards nighttime electricity use, but also home refrigerators and hot water heaters which, until the rise of the conservation movement in the 1970s, were phenomenally hungry appliances, and even today remain—with air-conditioning—the most electrically intensive items in a home.

Before he could set his sights on convincing regular folks to invest in this new array of electric appliances he had to first bring the city, with its traction companies, rich home owners, and manufacturers— all of whom were reasonably content with the private systems they had in place—into the fold of a single provider. To accomplish this, Insull made a series of interlocking business decisions. First, he radically lowered the price of electricity, making it competitive not only with costs relative to running a private plant, but also with gas. During the first five years that Insull managed Chicago Edison he halved the price for his central station electricity from 20¢ a kilowatt-hour to 10¢ (in 1897) and "proceeded every year or two to impose an additional 1-cent drop until prices reached 2.5¢ per kilowatt-hour in 1909." It was during the same period that the number of customers served by Chicago Edison skyrocketed into the hundreds of thousands.

In 1911, to ensure sufficient industrial load, Samuel Insull started selling "off-peak" power (the middle of the day and the middle of the night) to industrial customers at 0.5¢ per kilowatt-hour—a price that helped persuade factories and other industrial concerns running private plants to make the switch to grid-provided electricity. At this price, buying power from Insull was both cheaper in real money terms and cheaper when considering maintenance, upkeep, and necessary capital investment to keep private systems in good working order. For the utility itself, the reduced price of 0.5¢ represented a straightforward gain, because producing electricity was the literal least of their costs.

Of the many strange financial logics that adhere to the electricity business, two gave Insull the most pause. Both of these, it would turn out, had the same cause. First was that the lower the price he charged for electricity, the more money he made. Second, and related, was that his costs remained relatively constant regardless of how much electricity he sold. If he ran the Adams Street Station full bore only 5.5 percent of the time, the costs of running his company (including the station but also personnel, system maintenance, line upkeep, sales, and coal) was virtually the same as if he ran the power plant at full capacity 95 percent of the time. This was because, in Insull's own words: "By far the most serious problem of central station management and by far the greatest item of cost of the product is interest on investment." Thus, the more "units," or kilowatt-hours of electricity he sold, the more the interest cost per unit fell because this cost was being spread across a larger number of units. And the closer a plant's actual output was to its potential maximum the more the burden of this interest was offset by income. Every half cent they made selling power to industry was a half cent more than they would have made if their plant spent its middays idling for lack of load.

Like many of the logical developments Insull came up with to deal with the oddities of electrical power production and sale, this one remains with us today: industry still pays substantially less per kilowatt-hour than do residential and commercial customers, while relationships with industrial customers are privileged and cultivated in ways that seem, from the outside, to be fairly nonsensical.

To potential residential customers who were already making their own power, Insull made a similar argument, one grounded in good economic sense. Independence was more costly, almost foolishly so, than was subscription. Richard Munson explains: "To describe the disadvantages of small, on-site generators," Insull "often discussed the economics of powering a block of northeast Chicago homes. The 189 apartments, he explained, used an aggregate of 68.5 kilowatts of electricity for lighting. But because the lamps were lit at different times in different apartments, the block's maximum demand for power at any single moment was only 20 kilowatts. Therefore, Insull reasoned, a central power station providing 20 kilowatts would be more efficient and economical than a series of separate generating plants in each apartment with the aggregate capacity of 68.5 kilowatts."

Luring in customers with low prices and then using economies of scale and stories of modern progress to increase both the absolute number of people served and the amount of electricity each consumed formed the "grow" half of the "grow and build" strategy that for-profit utility companies would borrow from Insull and flourish under until the late 1960s. The other half was to build more, bigger, and more efficient power plants. And this was supported by Insull's greatest coup: he used government regulation to protect his interests from competition and to insure long-term, low-interest construction loans. As he courted big government, his business strategies were observed and slowly taken up by industrialists and monopolists elsewhere in the nation, who grew private electricity empires by the same means that had proved so successful for Insull in Chicago.

Southern California Edison, for one, took strong steps toward merging smaller, community power companies into its developing monopoly. Detroit Edison first took control of service in that city (which is now home to the least reliable grid in America), and then it slowly expanded into all of eastern Michigan. J. P. Morgan got into the game with the United Corporation, which managed electricity service in New York, New England, and much of the Southeast. Even Alaska got its monopolist as Wilber Foshay found his fortune in Steward's Folly, thirteen other states, the Western Provinces of Canada, and in Central America. All these men, with their top hats and cigar-clenched teeth, made private empires out of electricity by following Insull's lead until, by the closing years of the

1920s, ten holding companies controlled 75 percent of the American electricity industry.

To avoid being shut down by the antitrust laws then gaining in vigor it was necessary for all these early utilities to work hand-in-glove with government. As Insull was quick to learn, working with government was not the same thing as working with individual politicians or political parties—ephemeral in their real power and greedy in their pursuit of a permanent majority. Rather, a functional electrical infrastructure needed its echo in a permanent government bureaucracy. In this way one might limit real competition while also avoiding the uncertainties that came of colluding with particular politicians or parties.

As Hirsh discusses in detail in his history of what he calls the "utility consensus," the move toward government regulation as a survival strategy for the utilities had the good fortune of coinciding with the progressive politics of the 1910s. The main aim of the progressives was to rout out crooked politicians whose rule was grounded in relationships of patronage rather than systemized good government. Progressives and utility company managers thus found each in the other an ally, and what was born of their alliance was an agreement to monopoly; if the utilities accepted to be heavily regulated, the government at the state and federal levels agreed to grant them a guaranteed service area, within which no other electric utility would be issued a charter to function.

Unlike the trusts held by Standard Oil or U.S. Steel that attempted to control the entire U.S. market there were many electric companies—they simply never overlapped. In this way, the term "monopoly" took on a very particular meaning when speaking of power projects; despite the fact that there were numerous players in the market there was no competition between them because the market itself had been divided into quadrants, the borders of each enforced by a political apparatus designed largely for that purpose. In this the utilities bore a certain resemblance to modern-day street gangs whose territorial agreements, hashed out albeit with submachine guns rather than governmental subcommittees, sport a similar structure. One simply doesn't offer one's product in territory claimed by someone else. Prices are fixed by the provider and the structures of power in place brook no complaint.

Insull and his ilk thus grew and balanced their customer base though a mix of smart entanglements with government and the gobbling up of competitors. Five years after Insull arrived in Chicago he owned every other electric company in the Loop. He also owned all their generating stations, whole hosts of small power plants that were not in his opinion good for much. The era of small power was passing. The Adams Street Station, shuttered in 1894, a mere six years after it opened (it was simply too small for all the load Insull had recruited), was immediately replaced by the largest power plant in the world at Harrison Street. The Harrison Street Station was not only big in absolute terms but more efficient: its high-pressure condensing "marine" engines reduced the amount of coal needed to generate one kilowatt-hour from between ten and fourteen pounds to between three and five. As Harrison Street ballooned from 6,400 kilowatts in the early years to 16,400 kW in 1903 this efficiency translated into a difference of roughly 115,000 pounds of coal per hour. Though its newfangled engines used a prodigious amount of water, its location at the edge of the turgid Chicago River meant that this resource was—unlike fuel—effectively free.

The Harrison Street Station was the machine that in its grandeur and efficiency allowed Chicago Edison to begin reducing its prices in earnest. It also convinced "the Chief," as Insull was often called, of something he had only previously suspected. Plants could get bigger and they would get bigger, and the technological innovation that guaranteed their efficiency, could also be guaranteed to improve. In 1892 the Pearl Street Station ran at 2 percent efficiency—which is to say, it transformed about 2 percent of the potential energy in the coal it burned into electricity—twelve years later the Harrison Street Station had an efficiency rating of 12 percent and Fisk Street, Chicago's first AC power plant, completed in 1903, did even better. By 1940 even average power plants were running at 20 percent efficiency. One can see why Insull found technological improvement such a promising route toward profitability.

If Insull knew this to be true of his own era he suspected that it would always be true, that technological improvement would forever allow for greater reliability over a wider service area at lower cost. One could grow a central station, electric monopoly indefinitely.

In this he was not entirely wrong. For the next seventy years this is exactly how the electric utilities did grow while maintaining an exceptional capacity to produce a low-priced, infinitely available product. Until about 1969 there seemed no end to the amount of electricity America could use and still be well within the limits of what American utilities could provide.

When this logic at last broke it was devastating. Not the least to the cohort of electrical engineers who had never known anything but the truth of the grow (your customer base) and build (more and bigger plants). As the first waves of what would become the energy crises of the 1970s crested over the nation, the men who ran and regulated the business of electricity were uniquely unprepared for change. Their world had its natural laws, and when these laws quite suddenly ceased to hold true, they were thrown from their feet as if gravity itself had stopped pulling things to ground. Technological improvements would not, it turned out, increase plant efficiency forever, nor would individual Americans always choose lives that increased their electricity consumption. Lastly, the unexpected, drastic increase in fuel prices attendant on the oil embargoes of the 1970s, coupled with an equally surprising increase in the cost of building power plants—linked mostly to the vagaries of nuclear power—meant that for the first time since 1900, electricity prices began to rise rather than to fall.

The first blow came in the early sixties, when it became clear that technological improvements no longer promised increased plant efficiency; this was primarily a problem of physics. The second law of thermodynamics, and its corollary Carnot's theorem, dictate that temperature ratios limit the amount of work any given fuel can be expected to do in a heat engine. A traditional power plant is exactly such an engine: it turns fuel into heat. This heat is then used to convert water into a furious jet of steam directed at the blades of a turbine which, with their spinning, turn a shaft. This shaft then pokes into a giant electromagnet, and as the shaft spins inside the magnet, it produces an electric current. A system like this, that converts a fuel—any fuel, coal, plutonium, oil, gas, biomass, or trash—into heat, is going to top out at about 50 percent efficiency. This is an inevitable law. No power plant built by man or some yet-to-be-invented machine intelligence will ever do better than just under 50 percent.

There is another problem that limits the thermodynamic efficiency of a plant—the fragility of the metals from which boilers and turbines are forged. Technologically, we can now build a steam plant that works at 40 percent efficiency. We had some of these already in the 1960s. Practically speaking, this means superheating water to over 1,000 degrees Fahrenheit while upping the pressure to an awesome 3,200 pounds per square inch to convert this water straight into dry, unsaturated steam without boiling it. The construction of a machine robust enough to perform this task 24/7 for thirty years has turned out to be both fraught and expensive.

The closer a steam plant comes to 40 percent efficiency, the more routine maintenance it needs, the more often it breaks down despite the constant upkeep, and the more costly the high-tech alloys necessary to endure this onslaught. Sooner or later even these fancy crafted metals will fatigue and give way. By the mid-1960s it had become clear to utility men that a plant run at just over 30 percent efficiency was both the most reliable and the most cost-effective way to make electricity.

The truth of this has not changed in the fifty years since. In 2012, the best fossil fuel power plants in America ran at 42.5 percent efficiency—but this number is only for a few natural gas combustion (no-steam) turbines. The newest steam plants operating in the United States between 2007 and 2012, whether fueled by coal or plutonium or petroleum, came in right around 34 percent efficiency.

When critics of the way we make electricity seem to be speaking of wasteful practices when quoting such low numbers of plant efficiency, they rarely recognize that for a traditional power plant, 48 percent efficiency—not 100 percent—is the maximum achievable and that in fact, we made great strides during much of the twentieth century toward approaching this goal. That the end of this long asymptote toward perfect thermal efficiency arrived should have caught no one by surprise. The second law of thermodynamics has been known since 1824 and taught to electrical engineers since the profession emerged in the late 1880s. The belief that past successes might be projected indefinitely onto the future regardless of inconvenient truths like the workings of physics was a foolish way to run a business. This is true even of the electricity business, for which the bottom line is never actually at issue. Once they'd become government-regulated monopolies, profits for the utilities were guaranteed almost regardless of how they chose to run things. Every dollar spent promised a set return. What was shaken by this rather rude intrusion of Carnot's theorem into the business of making and selling electricity in America was that indefinite growth—as a business strategy—was no longer the only, nor even the best, way to proceed.

Nevertheless, seventy years is a long time during which lines on graphs and basic infrastructural logics did seem to follow a predictable, and largely positive, path. The truism that the future simply cannot be relied upon to mimic the past was lost as the third generation of electrical engineers entered the electricity business in the late 1950s. The consolidation of power that had propelled Samuel Insull forward into ever more impressive cycles of inventiveness had become the stagnation of a structure known to work profitably even when left well enough alone.

Unfortunately for utility companies, Carnot's theorem was not the only surprise that awaited them as dawn broke over the 1970s. They had been encouraged to switch from burning coal to burning oil in the 1950s and '60s—a costly retrofitting of power plants that was en vigor as the first OPEC oil embargo hit in 1973. The drastic fuel shortage that ensued saw prices rise more than 70 percent almost overnight and the notion of the shortage, as much as the reality of it, hit mainstream America like a sledgehammer. The skyrocketing price of oil affected not only how much it cost to make electricity in all the newly converted petroleum-burners but also for the nation's large unconverted fleet of coal-burning power plants. In 1973 any industry that had the capacity to use coal instead of oil was making the switch, a move that also drove the price of domestic coal through the roof. For the first time in their history, utilities had to raise the price they charged for electric power, shattering another seemingly natural law of the utility business.

Beneath the immediacy of this crisis for the utilities lay a slow-burning fire that would eventually unravel the economic surety of the grow-and-build years entirely—the burgeoning environmental movement. A newfound care for the natural world, while troublesome to the utilities, was less of a dagger to the kidney than the legislative and regulatory shifts that transformed this "care" into a viable political platform. Simultaneously, the shortages of the 1970s that pushed electricity prices up also produced a new relationship between Americans and their patterns of consumption. "Conservation" and "efficiency" became the watchwords of the 1970s. I remember being trained at school, in the middle of this decade, how to conserve electricity: the last kid out of the room always turned off the lights, the hallways were never lit, the heat was down in sweater territory, and the girls were encouraged to wear pants rather than tights under their dresses.

Daily practices of this kind drove down electricity consumption while environmental regulations drove up, way up, the cost of building power plants. The price of labor and material necessary for plant construction rose 120 percent between 1970 and 1979 (compared with 23 percent the previous decade), and time to completion of these plants also expanded. From start to finish, a power plant begun in the late 1960s took seven years to come online, while one started in the late 1950s had taken only five. Much of this increased time to completion was due to the incessant layering on of environmental standards after the Environmental Protection Agency was formed in 1970. Power plant emissions before this point were environmental horrors. Today, for all our worry about carbon emissions, we at least no longer suffer crop failures and fish die-offs from acid rain.

All of these developments combined to confound the economic truisms of power production: the price of electricity had always dropped; the cost of making electricity had also always dropped, while plant efficiency had always risen and electricity consumption too had always risen. That all of these factors changed at once was a series of blows to the utilities not at all unlike the final moments of a boxing match. Their surprise was that of the man about to hit the mat, caught so thoroughly off guard by a sudden volley of well-placed fists that he failed to protect himself against them, flailing a little bit more as each blow hit home.

Meanwhile, long an enterprise in the background of American popular consciousness, the hordes of protesters camping for weeks at a time around the perimeters of the nation's newest electricity-generating machines—the nuclear power plants—made the dull intricacies of everyday power production into front-page news. The utilities were being hit hard and their inability to respond with anything like aplomb was broadcast to the nation in prime time.

As unfortunate and confounding as these unforeseen developments were they hardly held a candle to what had actually broken. The seemingly natural law of grow-and-build, supported all along by a gentleman's agreement on the part of the utilities to be regulated and on the part of the government to regulate, in ways not too onerous to the utilities' profit margins, had begun to break down. This happened slowly with the Carter administration, and then much faster with Reagan-era deregulation.

More than any fact of physics or markets or culture, regulation in the form of long-term loan guarantees coupled with assured profit on investment had enabled electric companies to weather the intense infrastructural costs of building ever bigger, more powerful electricity factories and ever more expansive power networks. Regulation had also protected them from competition, ceding each absolute control over the production and distribution of electric power within given service areas. The reason for this long-standing state support was that electricity was deemed that peculiar kind of public good whose price is driven up rather than down by competition and whose availability is disturbed rather than assured by multiple, competing providers. In the 1970s, however, it slowly became clear that the supposed deleterious effect of competition was less a natural law than a bureaucratically enshrined functional reality. And like much of what Insull built, it had remained true for just long enough for its cultural roots to be forgotten.

The agreement to geographically delimit utilities and to allow the government to regulate them as monopolies is how we got the queerly speckled map of utility districts we have come to accept as normal. In most places, a single massive investor-owned utility holds sway over thousands of acres of territory and a customer base that can mount into the millions. You know these, you likely pay a bill to one of them every month. They have names like ConEd, National Grid, PG&E, PEPCO, Xcel, Entergy, and Southern Company. These are the direct descendants of the system that Insull built in Chicago, Miller built in Southern California, Morgan built in the Northeast, and Foshay built in Alaska and its neighboring provinces and states. There are 3,306 electrical utility companies in the United States, yet two thirds of us (68.5 percent) pay our bill to one of the 189 big for-profit, investor-owned leviathans.

The other roughly three thousand are tiny, nonprofit, municipally governed utilities or public utility districts (called a PUD in the West or a PPD in Nebraska) that provide power to a city or town. Los Angeles, California, has one; so do Clatskanie, Oregon, Anderson, Indiana, Coffeyville, Kansas, and Osceola, Arkansas, among two thousand others scattered across every state in the nation. Often these exist as islands in the midst of a vast sea of another, larger electric company's territory. For example, Burbank's utility is almost entirely surrounded by Los Angeles's utility, which is itself totally encompassed by Southern California Edison—the for-profit colossus of Southern California.

Tiny Rancho Cucamonga out in the middle of the California desert has a municipal utility of a mere four square miles, fifteen thousand daytime customers, 197 streetlights, and a single substation. It is surrounded for three hundred miles in every direction by Southern California Edison, which with 14 million customers over fifty thousand square miles actually has more employees than Rancho Cucamonga has people. Yet the latter is required to leave the former in peace. Rancho Cucamonga gets to settle its own problems, buy or make its own power, set its own prices (within regulatory limits), and provide for its own customers as it sees fit. No matter how delimited its service area or how small its customer base a municipal utility is just as much a monopoly as is an investor-owned utility. Both are protected by law from competition; both are prohibited by law from price gauging. And the only way to break out, or in, to another company's territory is through legislative action.

This is exactly what Marin County did in 2010. Many residents of Marin—a dry, half-empty, hyperwealthy, lefty stronghold in the sun-beaten hills north of San Francisco Bay—wanted the right to produce more of their power locally from renewable resources, or in their terms, to "locally curate" their energy while also managing their prices, rate structures, and customer relations for themselves. In essence, what they wanted was to get out from under the thumb of PG&E— the colossus of Northern California, which has never been known for having a light touch when it comes to community relations.

Though Marin had the legal right to break free from PG&E when a local vote authorized the switch, they found themselves in the midst of a much larger, indeed state-wide, battle with their former utility, an exceptionally wealthy, irate adversary. In addition to an admirably thorough public information campaign, PG&E sponsored a ballot measure that would change California law to require a yes vote by two thirds of all the residents in a community in order for that community to "aggregate" and then defect from their existing utility.

The wording is important. PG&E's measure sponsored not a two-thirds majority of the votes actually cast in an election—already a difficult task—but a two-thirds majority reckoned in terms of the total population of an aggregating area. Given that voter turnout in the United States rarely tops 60 percent, even in hotly contested races, the success of this ballot measure would have virtually ensured that no community in the state would ever manage to gather the necessary votes to secede from their utility. The right would have remained but the means of realizing that right rendered unattainable.

The careful placement of ads, editorials, and informational bulletins and the money spent by PG&E—a cool $43 million—made clear how desperately the utility wanted to keep the Pandora's box of community choice aggregation firmly shut. And when PG&E lost this ballot measure, despite having outspent Marin County 430 to 1 (Marin managed to raise slightly less than $100,000 for their part of the campaign), the borders of California's utility map were redrawn with a new player, Marin Clean Energy, standing guard over their own small patch of land within which they can decide what constitutes power.

This case is important not just because it did indeed open the door to community governance of electric power in California but because the tools used by PG&E to defend its ground were almost identical to those championed by Insull in the late nineteen-teens to ward off the public power projects then growing in popularity.

Samuel Insull didn't just build an infrastructure, he didn't just make a monopoly enterprise out of the most unlikely of candidates, he didn't just figure out how to fold government regulation into the very heart of a utility's finances, he also conceived of and pushed a dedicated propaganda machine, funded by customer rates, to ensure that investor-owned (for-profit) utilities remained the way the America made, distributed, and ran the business of electricity.

During his earliest years in Chicago Insull may have concentrated on consolidating adjacent electric power networks into a single, massive centralized machine that customers received power from and paid money into. But once this had been accomplished he turned his attentions to the preservation and multiplication of this system.

From the late nineteen-teens until the early days of the Great Depression Insull battened down the hatches and began to build walls. His goal was to render the privately owned, corporate way of making and distributing electric power unassailable. The battle Marin found itself in with PG&E was but a skirmish in a long-standing war between public power and investor-owned utilities. A war in which all the money and all the power money begets was on one side. Though curiously (as with Marin) this fact has been in no way a consistent guarantor of victory.

By the time the Depression hit, demolishing Insull's empire and a number of other utilities with rickety, indeed questionable, accounting practices that buried debt inside of holding companies, the issue was no longer one of private plants versus central stations but of public power versus corporate power. Nor was it a question of competing providers within a single locality, but of which sort of monopoly an electricity customer might find themselves a part of: a nonprofit municipal network or a for-profit investor-owned utility.

After the New Deal swung into effect in 1933 and complicated the scene with its big public works projects like the Tennessee Valley Authority (TVA) and the Columbia River dams run by the Bonneville Power Administration (BPA) and legislative interventions toward universal electrification (most notably, the Rural Electrification Act, or REA), the standard business models by which electricity was made, distributed, and sold opened to include rural cooperatives and federally managed power administrations. Despite this, all electric power remained monopoly-made, monopoly-distributed, and monopoly-billed, while all monopoly profits were government-regulated and all electricity prices were regulatorily determined. Last but not least, regardless of where one lived in the nation, rates were structured in such a way that the more electricity one used the less money one paid.

For example, in a 1934 bulletin, Southern California Edison details what their customers might expect to get from a modest monthly "light" bill. Saying: "In the typical six-room house, electricity does all of the washing, ironing and sweeping. It makes a pot of coffee and eight slices of toast every morning with waffles on Sunday. It supplies energy for radio three hours a day and provides for the occasional use of the curling iron, fan, warming pad, and other appliances—all for an average monthly bill of $1.92. For an additional $2.00 or $2.50 monthly, electric refrigeration may be added, while complete electric service including dishwasher, mangle, and a modern range to do all the cooking costs on the Edison market an average of only $6.55 a month. There is no cheaper servant on the market."

It is clear that this is not yet everybody's power: for most it is still a "light" bill, in part because few people had six-room homes in the mid-1930s, not to mention a spare $6.55 a month to spend on power for their household appliances (the equivalent of $110 in 2015). Nevertheless the future is clear, the home will become a hub of happy consumption, not just of electricity but of all the material conveniences electricity will make possible.

Growth was not so much an industry watchword as a dogma that would carry it, and us, forward until, bit by bit, in the late 1960s and early 1970s, all the truths of the electricity business began to break down. Only then, seventy years after Samuel Insull took the helm of tiny Chicago Edison, fifty years after he turned all of Chicagoland's electricity into a monopoly enterprise, thirty-five years after the collapse of his empire and thirty years after his own ignominious death in a Paris metro station, did the "natural" laws of the utility business, discovered and instrumentalized by Insull himself, prove to be little more than willfully held articles of faith and carefully engineered blindnesses.

In many ways it is more correct to say that Samuel Insull, and not Thomas Edison or Nikola Tesla or even George Westinghouse, made America's grid. He normalized and rendered profitable the central stations without which we would have no grid at all, just a bunch of factories, municipal buildings, and homes with little electricity plants in their basements. He imagined electric light and power as products for the masses not the few; he made it seem natural that the electricity business could only work as a monopoly, and he ushered in an era in which one of the most powerful things one could do with money, and to make money, was to use information to manipulate public opinion and influence public investment.

Small power, by the 1960s, was gone not only in fact, but it had been erased bureaucratically and legally. While sensible approaches to difficult problems, like providing all the electricity a growing America wanted whenever we wanted it, were limited in their scope by the formal structures and infrastructures of power that were built up in the wake of Samual Insull's remarkable takeover of Chicagoland's diversity of nineteenth-century electricity systems.

Samuel Insull's once insignificant Edison franchise set the stage for both the modern structure of electric power in America and also for a moral and normative sense of what it meant to be a good citizen-consumer. If in 1920 only the poor and the rural had no access to electricity, not just in Chicago but everywhere in America, then in 1950 it was inconceivable that power might be made privately on-site, and by 1970 only suspect radicals and known freaks were off the grid.

Americans were by definition people who had refrigerators, hot water heaters, air conditioners. We had electric lighting, wall outlets, and a multiplicity of things to plug into them. Increasing numbers of people, especially in the West, cooked on electric stoves and heated with electric radiators. We paid our bills in a timely manner, without much understanding how much electricity we were using, for what, and how much it actually cost. And though since the 1970s things have slowly begun to change, both culturally and in the business of electricity, for the most part these truisms still hold—we buy our power from a sole provider at a reasonable price, that we don't understand, and they insure that it stays on most of the time.

NOTABLE PASSINGS

On July 2, 1938, Samuel Insull stumbled and then collapsed in a Paris subway station. He was dead upon arrival at the hospital half an hour later. According to his wife his heart was rendered still by the task of mounting and descending such an innumerable quantity of steps. He was a poor man obliged to take public transit.

His obituary in the *Berkeley Daily Gazette* read:

> At the pinnacle of his power Samuel Insull sat on the directorates of 85 companies, was chairman of 65, president of 11, and through his holding companies controlled more than 6,000 power units, employing 72,000 workers in 324 steam, 196 hydro-electric generating plants, 89 gas plants, 328 ice plants. His companies served 1,700,000 customers, a population of 10,000,000. There were 600,000 security holders. The valuation of the system rose above the 3,000,000,000 dollar mark.

Commonwealth Edison, the final name of the Chicago utility built by Insull (and still its name today) was decimated by the stock market crash of 1929, taking the life savings of almost all those 600,000 securities holders with it. The Fisk Street Station, Chicago's first large-scale AC generating plant, was closed by labor unrest in 1942 and again for failure to live up to environmental standards in 2012. Making it, too, part of the story of how America changed around an electric power infrastructure too hardened in its ways to keep up.

CHAPTER 4

The Cardigan Path

Early in 1977 President Jimmy Carter, urged America, while sporting a surprisingly casual bit of beige knitwear, to turn down their thermostats. He'd been president for all of two weeks, and his position "fireside" for this little chat with the nation was no less symbolic than was his cardigan—the first ever seen on an American president.

For the better part of the decade the country had been in the midst of an energy crisis, and Carter had run, and won, on a platform of serious energy reform. We, as a nation, were going to need to do things differently. "We must face the fact," he said that cold February in his kindly Southern lilt, "that the energy shortage is permanent. There is no way we can solve it quickly, but if we all cooperate and make modest sacrifices, if we learn to live thriftily and remember the importance of helping our neighbors, then we can find ways to adjust, making our society more efficient and our own lives more enjoyable and productive."

In other words, cardigans, fireplaces, thermostats, and later, solar panels (thirty-two of which Carter would mount on the White House roof in 1979 in the midst of yet another crippling oil embargo)—these were to become watchwords for us all.

To modern ears, an executive call for a politics of thrift, moderate privation, cooperation between neighbors, and adjusting to (rather than mastering) circumstances sounds a bit tinny. So long have we been enjoined to consume our way out of economic crises, to buy more and spend more in order to stimulate manufactories to produce more and thus hire more, that the very notion that a sitting president might encourage us to do less of any of these things is kind of a shock. But Carter meant it. He wanted people to wear sweaters.

And people listened. We actually did start to consume less, not just during price spikes and the panics attendant to epochal shortages, but in a deeper and more abiding way conservation became a part of how Americans thought about, used, and managed energy.

This turn toward conservation and energy efficiency was the first crisis, of three, that would shock the electric utilities during the Carter era. Their business model, long premised on endlessly increasing consumption linked to endlessly increasing production of ever cheaper electric power tied, in turn, to the construction of ever bigger, more technologically advanced power plants, had begun to fail. For the first time in the storied history of electricity, consumption dropped; for the first time their power plants grew too costly to complete; and then, much to their surprise, something else happened: monopoly control of the grid was wrested from them by legislative fiat.

This was initially done in such a minor way that most utilities and many politicians (including some who had voted the change into law) didn't much notice when it happened. The men who wrote the legislation, the men who voted for it, and the lobbyists of utility companies missed the reinstatement of a minor sub-clause of a sub-act of an omnibus bill called the National Energy Act that only just squeezed through Congress after significant cajoling and administrative arm twisting. It passed in the House by a single vote (207–206) in the autumn of 1978.

What this clause said was simply that the utilities would need to buy, and move to market, electricity produced by any facility with an output of less than 80 MW (about a tenth of what might have been produced by an average nuclear power plant at the time). And, just as important, they would be obliged to pay a nonmiserly rate for it. This rate would be set according to what were called "avoided costs"—that is, the money it would have cost the utility to make that precise amount of electricity for themselves.

Even though few realized the implications at the time, section 210 of the Public Utilities Regulatory Policies Act (PURPA for short) effectively broke the utility's total control over everything that entered, moved through, and exited their power system. And it worked, not just because of what it mandated but because of the vehemence with which this mandate was taken up and implemented. Americans were fed up with the powers that were, and grasping at every straw made available to them (including the election of Jimmy Carter himself), we changed the power industry, very slowly, but inexorably and forever.

One reason this subsection was little noticed or troubled over at the time was that no one could have predicted the fervor with which it would be enforced, another reason, however, for its having been overlooked was the ambitious scope of the National Energy Act as a whole.

In a 2012 address to the Senate Foreign Relations Committee Meeting, President Carter recalled that the act "put heavy penalties on gas-guzzling automobiles; forced electric utility companies to encourage reduced consumption; mandated insulated buildings and efficient electric motors and heavy appliances; promoted gasohol production and carpooling; decontrolled natural gas prices at a rate of 10 percent per year; promoted solar, wind, geothermal, and water power; permitted the feeding of locally generated electricity into utility grids; and regulated strip mining and leasing of offshore drilling sites. We were also able to improve efficiency by deregulating our air, rail, and trucking transportation systems."

Given the litany of changes enacted by this single piece of legislation one can see why permitting "the feeding of locally generated
electricity into utility grids" might initially have slipped past a radar
tuned to the measure of significant things. All of the copious features
of this bill had an impact on the ways in which America used, produced,
managed, and imagined energy. In this, as with his other bureaucra-
tizing and legislative attempts to rationalize America's relationship to
power, the President got it right. At the end of his lone term in office
we had our nation's first ever National Energy Plan (the National
Energy Act was its legislative form), a Department of Energy instead
of the more than fifty unrelated government offices previously charged
with overseeing national energy policy, a Strategic Petroleum Reserve
to help smooth wildly vacillating oil prices caused by our national
dependence on foreign oil, and the National Renewable Energy
Laboratory, whose mandate was to explore and make feasible renewable and small, decentralized power options.

In order to understand how such a seemingly minor encumbrance
to utilities rose with time to the level of a crisis that in the present
is often, if somewhat melodramatically, referred to as a "death spiral,"
it is necessary to spend a moment in the 1970s, a time in which we
as Americans were becoming deeply aware of, and increasingly
concerned about, the vulnerability and incomprehensibility of the
systems that sustained us.

We, as a country, had spent decades, indeed most of the twentieth
century, and billions of BTUs in the project of national uplift. By 1970
we had largely succeeded in this task. America was a wealthy nation
and an infrastructurally sound one. Yet, the system we had devised to
make us rich had also developed some obvious downsides. The
government had gotten us ass over teakettle in Vietnam, and at home

our prosperity was made fragile by our dependence on foreign oil, the steady supply of which we couldn't control, while the material bounty that flowed from our industry had saddled us with a set of pressing environmental issues that regular folks had no power to correct. The institutions that ran America in 1970 were making decisions that increasing numbers of people didn't agree with, and yet these power players seemed beyond the reach of anyone to affect. The protest culture of the late 1960s was in part a response to the inaccessibility and calcification of state and corporate powers, but it was also linked to a growing understanding that the steady stream of engineered goods these powers provided did not come without cost. Pollution, like shortage, was becoming part of the national consciousness. Certainly not everyone was jostled into awareness in the same way, but enough Americans had become interested in systemic change that Carter, an unusual pick, was elected to the highest office in the land.

By the time Carter took office, every home in America was its own miraculous technological node, built into a complexly woven support net of wires and pipes and ductwork. By 1976 everyone in America who wanted it had electricity, indoor plumbing, central heating, a refrigerator, and a phone. Our workplaces were similarly well served.

Living in these homes and laboring in these workplaces changed us. It only took a generation after the end of the Depression for Americans to become consummately modern individuals, until as a nation we had lost working knowledge of a coal brazier, a kerosene lamp, a latrine, an ice box, a well, a mangler, or anything else more complicated than a switch, a button, an outlet, a socket, a tap, or a flusher. And yet, it was also the case that almost no one had any idea how the replacement technology (a coal-burning power plant for a brazier, or sewer treatment plant for an outhouse, or water purification plant for a well) worked. By the mid-1970s, who, driving down the highway, could tell the difference between an oil refinery and a coal-burning power plant? A sewage treatment plant and a water purification plant (or, for that matter, a fish hatchery)? A telephone wire, a TV wire, and an electricity wire? Realistically, who needed to be able to distinguish among these classes of support? Nobody, except the professionals charged with maintaining them. As long as the switches, buttons, outlets, sockets, taps, and flushers worked, the benefit of all the rest having become distant undertakings running along long wires and through long pipes was that they were no longer our immediate concern.

Thus did America lose one kind of knowing—that involved in managing a low-tech household—without gaining another kind of knowing—that of the distant complexity undergirding a high-tech household. High-tech here doesn't mean digital, though that trans-formation was also on the horizon by the late 1970s; rather, it means a modest but orderly entanglement of analog systems that provided for a comfortable life. These intimate infrastructures freed us up to live in ways unrecognizable to even big dreamers two generations earlier; our "normal" was blind luxury from the perspective of almost any other time in human history. One result of this was that we got to think about other things, do other things, and live longer, less disease-riddled lives. The problem was that all of these systems beyond the scope of daily understanding had consequences, and by the 1970s these consequences had become almost impossible to ignore.

The oil embargoes of 1973 and 1978 brought a nascent under-standing to the nation as a whole of the global complexity and vulnerability of our gas tanks' supply chain. The smog crisis in Los Angeles, which began in the 1950s and lasted through the 1960s, made the link between machines—including coal-burning power plants—and their effects palpable to Southern Californians' lungs and red, itchy eyes. Love Canal in 1978 in tiny Niagara Falls changed pollution from an abstract evil to an immediate horror show.

Though environmentalism grew in force and impact with Love Canal and similar "Superfund" sites, the partial meltdown of one of the nuclear power plants at Three Mile Island, Pennsylvania, in 1979 shifted the tide for good. This event seemed to prove to nuclear power's naysayers that they had been right all along. Nuclear was a no less poisonous way to make electricity than was coal that blackened lungs and smudged pores, and oil that was uncontrollable because of its origins on the far side of the planet. More than this, each mess, from DDT to acid rain, seemed to further convince a generation already angered and radicalized by the excesses of the Vietnam War that without an abiding wide-scale commitment to fundamental change disaster did indeed loom for America (and maybe, some suspected, we even deserved it).

By the late 1970s the problem was not so much awareness of what was going wrong, but rather how to change any of it. The systems in place that made smog, lakes of industrial pollutants, sweet light crude, and war had been hardening, in some cases to the point of total calcification, since the earliest years of industrial consolidation. In this the electricity business was no different: there was no way for regular people developing a new environmental consciousness to have a say in how power was made, from what, and with what after-effects except through protesting or opting out. Many Americans did both. Protests around nuclear power stations blossomed throughout the late 1960s and into the 1970s as thousands camped out around these power factories whenever they were due to open. Many more people were dissatisfied without taking to streets or going off the grid entirely. The drop in electricity consumption nationally was one sign of this discontent. Even people who couldn't afford to "turn on, tune in, and drop out" could follow the Cardigan Path, and so they did.

Despite these developments, the industry's stranglehold on generation meant that there was no alternate route that different logics of power production might follow. There was no place for choice or self-sufficiency and no reward for conservation or efficiency. Carter may have championed less rather than more, but before his legislation began to force change upon the status quo, acts of conservation or efficiency were without reward. One could use less power, but because of the tiered rate structures that charged the most for the first few kilowatt-hours used in any given billing period—a system put into place by Samuel Insull—these efforts were only minimally reflected in the dollar amount owed on the bill.

The impossibility of transforming a budding environmental consciousness into effective action against entrenched industries was not just an issue for individuals, but also for states and other communities attempting to enact their own responses to environmental degradation. California, for instance, had passed a bill into law, in the 1960s, that would mitigate pollution and encourage renewable forms of electricity production but could not enforce it because of the ways in which federal regulations protected big polluters from attempts at state reform. America was in a straightjacket and her frustration created a particularly fertile environment for even tiny seeds of hope, like that provided by section 210 of the Public Utilities Regulatory Policies Act. The Carter administration may have given the act its words, but the spirit of the times gave it its claws.

Nor was it just that the utilities had grown blind to the people; they were also isolated from governmental whimsy and reform by virtue of having their own dedicated regulatory bodies. The very institutions that had protected them initially from competition had become the agents of their cloister, divorcing the utilities from the effects of the market on their business decisions because they were guaranteed a modest return on every dollar spent. Buy a new office chair for the accountant? Earn a dime. Build a new coal-burning power plant—earn a hundred thousand dimes.

By the 1970s the utilities had ceased to live and function in the real world. Their rules were not our own. Their power had grown absolute, plodding, and blind. Energy historian Richard Hirsh takes it one step further, arguing that for decades the utilities had been attracting the bottom of the graduating classes from engineering schools: the students who didn't want an exciting career in "the glamor industries—electronics, aerospace or computers" or who weren't quite agile enough to land a more interesting job. It was a stagnant sector that promised no adventure and a steady paycheck. As a result, the most risk averse and least facile minds were running the game.

One of the reasons that section 210 of PURPA made it through the legislative process and into the law was that it didn't at first glance challenge any of this. It left the utilities as monopolies untouched and instead reformed an almost accidental right they had been given all the way back in the early 1900s; it stripped them of the right to control the market for electricity not as sellers but rather as buyers. PURPA actually left their monopoly powers well enough alone, but summarily retracted their powers as monopsonies.

Most people know what a monopoly is—an entity that maintains control over the supply of a good or service, like how U.S. Steel controlled the market in rebar in 1900 or how the Soviet State controlled the availability of pretty much everything from toothbrushes to potato sacks within the vast, sprawling territory of the Soviet Union. What is less common knowledge is that the monopoly has an echo on the consumer side of the business world called a "monopsony," which is the sole customer for a product. We don't talk about these much in America because it is so rarely a problem. Here consumers are such a precious part of commodity systems that even companies eager to control a market as a provider of a product or service (a monopoly) will rarely so much as flirt with the notion of controlling a market by being its sole customer (a monopsony). Nevertheless, a monopsony is every bit as much a kingmaker as a monopoly—and America's utilities before PURPA were a rare instance of both. Not only were they the only ones providing electricity within a given territory but they were legally also the only buyer available should some other intrepid industry—or one that generated a lot of excess heat in its daily operations, like a smelter—try to make and sell electricity within this same territory.

The utilities managed the grid, they made the power, they owned the wires, they distributed the electricity, and they collected the money. They were also regulated in such a way that anyone else with a notion to sell electricity couldn't. The law prevented other electricity makers (by dint of not providing a license) from building their own distribution networks and entering into competition with the existing utility. While utilities, for their part, either refused to buy power from these third parties outright or they set such a low price per kilowatt-hour that most wannabe power plants, when they did the math, realized they would be losing money on the project and opted out. This was part of the reason there was no "alternative" generation during the middle years of the twentieth century. Even power systems that had existed well before the utility monopolies had been established, such as small hydro in streams (popular in California well into the 1920s) or cogeneration plants (which in 1908 produced almost 60 percent of the electricity made in the United States), had been almost utterly erased by the utilities' preference for large, dedicated power plants. And since no one else could enter the market, because the utilities wouldn't buy their product no matter how cheaply or cleanly they made it, that was pretty much that.

Regulation, back when it was hashed out in the early 1900s, had presumed exactly this kind of vertically integrated electricity system. No one really anticipated that one day more than half a century later that same entity might be in the position of a "customer" for modest quantities of electricity produced by someone else. Monopsony powers had been granted to the utilities almost by accident.

This was what PURPA reversed: The utilities could still be monopolies, but they couldn't be monopsonies anymore. The utilities now had to buy power from entities making small amounts of electricity in their territory; and they had to pay the same rate for this independently produced power as it would have cost them to make it themselves. This second clause notably used the market to force small power producers to be more cost effective than the utility. To make any money they had to make electricity for less than the utility's "avoided costs"—a fuzzy term that could be (and was) interpreted to include everything from the cost of fuel to the amount of new generation a utility *wouldn't* have to build because of the electricity they were newly obligated to buy from other sources. PURPA was, in other words, a smart bit of anti-monopsony regulation.

So long regulated as monopolies, the utilities were not initially alarmed that they would now be regulated as monopsonies as well. This doesn't mean they were exactly happy about the terms, it's just that they were almost myopically concerned with losing their ability to offer so-called promotional rates.

By the late 1970s small had become beautiful and less had become more. There was no place in this new world of modest, consensual privation for rate structures designed to enjole people into using more power than they actually needed. Regardless, promotional rates were also one of the few tricks left in the utilities' toolbox for keeping the cost of their massive investments in infrastructure from dragging them into the red. From the last chapter you'll recall that the biggest problem with making a profit from electricity is that a power plant only pays off if it is used at close to maximum capacity twenty-four hours a day. Reduced consumption, whatever the cause, was a distinct threat to the utilities' main way of making money. Once everyone had electric power there was no real way to grow a customer base; the only option was to manipulate how much power these customers used and when. Promotional rates were the utilities' best way of doing this.

This tried-and-true means of convincing customers, most especially energy-intensive industries, to increase how much electricity they used was by offering special deals like almost free power in the middle of the night or half-price power after the first 60 kilowatt-hours used in any given billing cycle. Always an important part of their business model, such offers were critical in the 1970s as consumption was falling overall and power plant construction was reaching its apex. This effective means of maintaining an equilibrium in the electricity industry was also under threat from PURPA, and the utilities, as a result, were more worried about whole sections of the Energy Policy Act related to rate regulations than they were about buying a little bit of power from a scattering of tiny producers.

Thus, though the utilities did lobby against section 210 of the act—at one point they even managed to cut it entirely—nobody paid much mind when it was slipped back in by New Hampshire senator John Durkin. Senator Durkin's goal was to assure that a garbage-burning plant in his district could start to move the electricity it was making to market in the Boston area (and maybe even increase their capacity by starting to burn wood waste from New England's forests, what we now call "biomass"). The utilities thought they had bigger fish to fry than fighting Congress about the rights of a trash burner to sell a little power. In this, as it turned out, they were wrong.

In the United States in the 1970s there were still two types of electricity producers that were not part of the national grid. First were the cogeneration plants, like the New Hampshire trash immolation facility. These produced so much excess heat that running their steam through a small electric generator had no appreciable effect on the factory's other tasks. Second were the hippies and other do-it-yourself small power innovators. By the mid-1970s both groups were already making their own electricity, though both were infinitesimally minor players in the national electricity game. In 1975, cogeneration plants were producing about 3 percent of the nation's power and selling none of it, while what the off-the-grid folks were up to wasn't even worth the trouble of quantifying. Nevertheless, in the 1970s, minor players were the only real direction one could turn in hopes of breaking the stranglehold the major players had over almost everything. In crafting section 210, then, the Carter administration thought it would be wise to integrate the one (cogeneration) and politic (and perhaps interesting over the long run) to include the other (hippie power). In this they were not far wrong.

If so-called "frontier electric technologies" such as small hydroelectric dams, solar electric systems, and wind power were included in the bill as a means of supporting innovation at the smallest scale, more of an experiment than anything else, cogeneration was neither a new idea nor a new technology. From the earliest days of electric power, factories had been making electricity, at first only for themselves, but as the grid had grown, this almost incidental current had initially been included in the national mix of generations. More electricity was being produced in the United States by cogeneration than by utilities as late as 1912. But by 1962, all that had changed—industrial cogeneration plants were, at the time, delivering less than 10 percent of the nation's electricity. By 1978 that number had shrunk to 3.2 percent. Without direct government intervention, the number seemed destined to fall all the way down to nothing at all.

As it slowly became clear during the early 1980s that PURPA would stick, cogeneration began its climb back into the double digits. As of 2015, there were more than 3,600 factories in the United States making 12 percent of our electricity as a by-product of their own industrial processes. The U.S. Department of Energy's current goal is that 20 percent of America's electricity come from cogeneration plants by 2030. If its nesting efficiencies suited the prevailing attitudes of the

1970s, cogeneration also maintains a solid fan base among electrical engineers because its double use of the same heat increases plant efficiency to over 50 percent, leaving Carnot's theorem in the dust. Existing means for increasing power plant efficiency cannot be made to do better than cogeneration does by its very nature.

No longer just a question of a single New Hampshire trash burner too close to Boston to regret all the dollars it's losing as it vents its waste heat rather than recycling it into electricity, cogeneration today has officially entered the mainstream. So, too, have the chancier technologies—wind and solar most especially—though their path toward ubiquity was much less sure.

Though the integration of cogeneration into the national grid was clearly wise, PURPA also made it explicit that alternative renewable power projects would also be granted the right to sell their power to the local utility. As long as these generators remained smaller than 80 megawatts and produced electricity for less than the avoided costs of the utility then they would, just like the cogeneration plants, be guaranteed a buyer and a fair price for their electric power.

While PURPA was wending its way through the courts—a process that lasted five years—small power producers began taking out loans, buying land, fixing up old river dams, and erecting wind turbines. Nowhere was this more true than in California, which had been preparing, institutionally, for just such exigency for more than a decade. In 1983, when the Supreme Court confirmed the legality of PURPA and put an end to further legal dillydallying, California was ready. It had a host of small energy entrepreneurs with almost operational projects and it had local legislation in place to support electricity production that was verifiably environmentally friendly— what we would now call "green." A surprising number of permits for small-power construction were filed almost immediately, 1,800 for dams alone, as well more than 16,000 wind turbines and even some

for experimental solar projects. And all this new small-power generation was scattered to the four winds—in the river canyons and old
mines of the Sierra Nevadas, in the desert outside L.A., in the passes
between San Francisco Bay and the Central Valley, even in downtown Sacramento, and almost all of it was variable. It blew with the
wind, fell with the water, and warmed with the sun.

The utilities quickly found themselves with a plethora of new
problems. Never before had they had to deal with variable generation, never before had they had to deal with distributed generation,
and never in the seventy years of their existence had they lost control
over the production side of their business. At issue wasn't that they
suddenly had to integrate a massive amount of new power but that
they weren't getting to decide how much, where, or when relatively
small amounts of electricity would come streaming onto their power
lines. They just had to pay for it and distribute it when it got there.
Their business model had no provisions for this new reality.

It demanded that they behave less like Soviets.

For half a century the daily, quarterly, and annual running of the
grid had been accomplished in meetings between the managers of
various arms of a utility. Notes were taken, plans made, and later
operationalized. In place of this centralized and somewhat perfunc-
tory organizational routine the utilities now needed something
like real-time flexibility. Even if the scale was initially modest, the
introduction of variable, distributed generation for which they would
have to pay a market price was, for the utilities, a little like aliens
coming down from outer space and asking them to enter into an
intergalactic energy alliance. It's fine in theory, but unimaginable in
its details.

Take a simple seeming thing like paying market price.

The utilities weren't exactly clear by the late 1970s how much it
actually cost to make a kilowatt-hour of electricity. Much like any
monopoly, they had not needed to take the market price for their

product seriously into account. For fifty years they had been free from competition, they had had the price of their sole product set for them by regulators, and they earned a return on what they spent regardless of how frivolously. Over this same period their business model had become wholly reliant on guaranteed low-cost loans for big-budget items (like nuclear power plants). As a result they were more attentive to shifting political tides than to the market. By the time PURPA rolled around there was no longer anyone working in the electricity business who remembered what it had been like before, when competition was part of what governed the winners and losers of the electricity game. In 1978 they could hardly account for uncertain fuel prices, uncertain consumption practices, and the mysterious vagaries of environmental regulation. The problem of accurately estimating cost, profits, and price only became exponentially more difficult when applied to the future, which PURPA was now asking them to predict.

Meanwhile, there was some fear on the part of the Federal Energy Regulatory Commission (or FERC, founded in 1977) that the utilities would take the project of doing the math on the single "avoided costs" parameter of the transition as a means to stall implementation of the entire bill indefinitely. If each small power producer was treated to a long bureaucratic process before receiving a viable contract, then PURPA would fail; the letter of the law having become a tool in the demolition of its spirit.

To reduce wrangling and confusion and to make certain that PURPA was in fact implemented, FERC mandated that utilities, or in certain cases states, do the math thoroughly, but only once, and then use the number they found to determine all initial contracts with non-utility electricity providers. So labor-intensive was this task that the joke at the time was that PURPA ought to be called the "Full Employment Act for Economists of 1978." Even this massive deployment of people good at figures yielded unsatisfactory results since, as then president of the California Public Utilities Commission (and later U.S. secretary of commerce) John Bryson pointed out in 1982, accuracy "will vary with the time of day, season, and term of contracts, as well as the utility's marginal fuel or next planned facility." The level of flexibility that the utilities found themselves needing, most especially in their attempts to link pricing models with time-of-day rates, would not be realized until the Internet would bring real-time arbitrage to electricity markets almost fifteen years in the future.

The reason the price per kilowatt-hour had to be set at the time of the initial contract, even if that contract stretched out for fifteen or thirty years, is that small power providers needed to know how much they were going to be paid for the electricity they would be selling, not just when they brought their projects online, but for the foreseeable life of their infrastructure. In order to win a contract they had to first demonstrate that they could produce electricity more cheaply than could the local utility. And second, they, too, like large power providers, stumbled over the costs of facility construction. Almost all the new players in the generation game had a significant outlay of capital at the very beginning of the process, which they were required to tackle *without* the low-cost-guaranteed loans that had made traditional utilities big builders since the Insull era. And though these same small producers were unencumbered by the regulations that had made traditional utilities slow to innovate, they still needed to project a guaranteed income to attract investors at the very start.

Once the initial requirements of figuring "avoided costs" were met by state regulatory agencies, often in conversation with the utilities, went about "offering" contracts in very different ways. In some states, such as Vermont and New York, the same price per kilowatt-hour was offered to any and all new electricity providers regardless of the utility doing the buying; in New York this was called the "six-cent" law as it paid out a flat, invariable, 6¢ per kWh for all early PURPA era contracts. In other states, like Virginia, non-utility providers bid for contracts. For example, in 1986, when Virginia Power wanted to add 1,000 kilowatt hours of generation, for use by 1990, they held an open auction, received 5,000 kilowatt hours' worth of bids from fifty-three different companies from which they selected seven that promised 1,178 new kWh. This system of competitive bidding would become the most revolutionary of the many variants tried out by different states as it allowed for something like free-market competition to actually, and at long last, enter the governance of the grid. Once awarded the contract, these new, small power entrepreneurs "built out" the promised generation facilities, and on the appointed date, they began contributing a predictable kilowattage to the grid.

Not all states followed this model. Some avoided the bidding process and, rather than producing a single standardized contract, produced a series of them. California is the state that lives in the history books because of the unique (some might say spectacular) way in which their contracts opened the states' electricity markets to "frontier electric technologies" rather than to cogeneration, which was the winner in most other places. California gave the hippies their due, and many of these leapt at the chance to make the state's electricity, often with far more gumption than manifest skill.

In California there were four standardized offers, of slightly different lengths and with slightly different terms, which once written were made available to cogenerators and other wannabe small power providers to choose between. The first three were all a little too beneficial to the utility side of the contracting parties and as such received such a lackluster reaction that a fourth, interim choice—called by the dashing name Interim Standard Offer #4, or ISO4—was made available. It had terms remarkable enough (and it turns out, foolish enough) that not only did a great many small power producers sign up during the eighteen months between 1983 and 1985 that it was available, but a much larger percentage of these were "alternative" renewable projects than in other states. Unlike back east, for example, where in some states upwards of 90 percent of the new generation contracted during the first decade post-PURPA came from cogeneration, in California it would be wind that stole the show.

"The combination of federal and California incentives and innovative state regulations launched the wind industry in the U.S.," write wind advocates Randall Swisher and Kevin Porter, and these "gave California the short lived title of having the most installed wind capacity in the world. Grid integrated wind development increased from 10 MW in 1981 statewide to a cumulative total of 1039 MW in 1985 [the equivalent of two coal-burning plants]. By the early 1990s as construction was completed on most of the remaining ISO4 contracts California had about 1700 MW of wind projects in place." All in 80 MW chunks. "By 1990, California had become home of 85% of the world's capacity of electricity powered by the wind and 95% of the world's solar power electricity."

The ISOs that helped structure the early days of PURPA's implementation in California functioned a lot like Medicare does today. In the case of Medicare, different retirees have different needs; some need good prescription coverage, others need lots of very cheap doctor visits, others need regular high tech services (such as dialysis). They then look to the small print of the plans (contracts, in fact) on offer and sign the one that seems to be the best fit. Every five years or so they get to revise this decision and choose again, though usually from a slightly different set of standardized contracts. It's not free-market competition, exactly, but it means that individual cases can be "batched" while giving the minority partner in the interaction the ability to pick. No single contract is a perfect fit, but the reality of choice makes compromising on what are in fact non-negotiable terms a less bitter pill to swallow.

ISO4 was unique among California's offerings in that for the first ten years of the contract it paid more than the avoided costs to the

GRETCHEN BAKKE

utility; it then paid less than this amount in the final years. This was particularly appealing to entrepreneurs because of how expensive it was to build new energy infrastructure. Unlike cogeneration, which had been around in one form or another for a long time, there was no tried-and-true means of making wind power or solar power in the early 1980s. Even the many small hydro projects that blossomed across California during the first post-PURPA years, which used known technology, still involved building new dams. All of this necessitated a lot of capital up front. The ISOs, and most especially number four, provided a solid assurance of income over the long term. This in turn made wind, solar, and small hydro quite suddenly and surprisingly into good investments.

In the early 1980s Tyrone Cashman sat at the "Wind" desk of California's aptly named Office of Appropriate Technology in Sacramento. He is largely credited with bringing wind power to California, because he understood that entrepreneurial will alone, not even when coupled with utilities' signatures on the right pieces of paper, would be enough to bring about a renewable energy revolution (to which he was, and remains, deeply committed). Tax credits, Cashman figured, were the answer.

The Federal Government, as a part of PURPA, was already offering 10 percent credit for investment in non-utility generation, tuned up to 15 percent by the Crude Oil Windfall Profits Tax Act in 1980, and further sweetened by an additional 10 percent investment tax credit which could be applied to any capital investment "regardless of whether it was used for a robot at an auto plant in Indiana, a giant shovel at a strip mine in Ohio, or a wind turbine in California." A 25 percent tax credit provided by the Feds for investing in the electricity generation business was already a very enticing deal to

investors. What Cashman did was double it. In California, any investment in renewable infrastructure (including solar and wind power, but
also solar hot water heaters and other non-electricity producing uses of
renewables) was rewarded with a tax break of nearly 50 percent.
Money poured into the state, and most of it was transformed into the
steel stalks and massive rotating blades of wind turbines. There
were veritable forests of them.

What nobody involved with crafting ISO4 or the tax credits that
accompanied it had expected was how well it all might go. Doing
anything new in relationship to the bulk energy game had for so long
been impossible that the best these new energy entrepreneurs and
ideologues had hoped for was to make a tiny dent in the existing
edifice. What they got instead was a party that everybody (even
people they hadn't exactly invited) came to. Taken together, ISO4
and the state's overly generous tax credits created a glut in the renewables market. And as oil and natural gas prices plummeted in the
mid-1980s, California's two big utilities were quite suddenly being
required to purchase more power than they needed for far more
money than they could recoup through their rates. Meanwhile, the
Reagan-era deregulation of the savings and loans in 1986 took
Cashman, and most everyone else, totally by surprise. Suddenly
everyone, not just the superrich—"the baseball players, football
players, plastic surgeons and Southern California actors" that
Cashman's turbine investment plan had been meant to attract—could
secure the financing necessary to buy their very own personal wind
turbine. Far too many people, in his opinion, did just that.

"There just weren't any stodgy bankers out there," he explained to
me more than a quarter of a century later, in his modest Marin County
living room. "There were only wildcat bankers, who weren't going to
lose a thing no matter what they invested in." And then "all these Wall
Street types turned up, hiding themselves so we wouldn't see them,
but crawling all over this place." This sentiment of too much

investment too fast is echoed by energy analyst Paul Gipe, who points out that the "tax credits were so lucrative that they attracted those who knew more about constructing a deal than about building wind turbines." Unwittingly Ronald Reagan had created one of the weirdest marriages American business has ever seen, as Manhattan investment bankers scrambled to buy up wind turbines made by commune-living, Vietnam-era draft dodgers. The result was that California, by the mid-1980s, had a massive wind bubble, ripe for popping.

California might have been the planetary center of wind energy in the mid-1980s, but their turbines were more machines for churning out visions of greener futures than actual watts. The buy-yourself-a-wind-turbine plan had become so appealing that, in Cashman's words, "an awful lot of machines were put up that were worthless." Nobody, at least no one in America, had figured out how to build an industrial grade wind turbine. What we'd figured out instead was how to finance them. The state-of-the-art turbines during those early heady days were, in Cashman's words, "prototypes" that, unfortunately for California, operated according to the wrong logic.

America's first turbine engineers were aeronautical engineers who had opted out of working for Vietnam-era helicopter companies. As a result they designed their turbines with floppy flexible blades based on the aerodynamics of helicopters. It turns out that the blades you need on a helicopter are the exact opposite of the ones that make for a successful wind turbine.

"On a helicopter," Cashman explains, "you need lightness because you have to get off the ground, but you don't want lightness in a wind turbine. Heavy is what you want, brute force, and another thing—when a helicopter goes in the air it has a chance to move with big movements of wind whereas a turbine can't . . . it's just standing there beating its brains out against the wind all the time . . . and there were other problems too: flexibility of joints and everything; wind turbines don't need that either." Since the helicopter guys had the wrong theory, once they got to the trial and error phase of things they never got very good results without really understanding why. Their machines couldn't be tinkered into efficacy.

By 1989, a decade after the passage of PURPA, the average time a California wind turbine worked at capacity before breaking down and needing repair was seven hours. California had thousands upon thousands of these and none of them really worked.

"We kept saying a turbine lasts twenty years. That was our vision." Cashman laughs. "But none of our turbines lasted twenty years. As a matter of fact, Alcoa Aluminum had a huge Darrieus Rotor that they had made, just straight aluminum. They thought they were going to be the wind energy guys, you know, really big. It fell down the day before we had the annual American Wind Energy Association Conference at their headquarters. It was going to be the big symbol, and it fell down—it got resonant vibrations."

I can't tell at this point in the story whether Cashman finds all of this tragic or funny. After all, California's success in building wind power in the 1980s was critical in establishing the idea that renewables could be used to generate electricity for power grids at a scale comparable to fossil-fuel-powered plants. In 1984, according to Cashman, even with the questionable life span of their machines, California's nascent wind industry produced as much electricity as San Francisco used that year. They had proved that even with prototypes they could make enough energy out of thin air to power a major American city. This was a real victory.

If American war defectors had been the only ones trying to figure out how to make electricity with wind, PURPA, ISO4, and some of the most generous tax credits in history couldn't have saved it. The industry would have crashed and burned; it almost did. But as it turned out, the American hippies weren't the only ones turning their sights to the wind. The Danish hippies, with their small, flat, wind-swept country, had the same idea, and if Americans were mostly trained as engineers, the Danes were former blacksmiths. "They had a totally different relationship with metal," Cashman explains. "They spent their time fixing large-scale farm equipment, so machinery was their model."

"Our first and second generation of prototypes, still needed . . . reconceiving," Cashman says, trying to put it nicely. "Whereas their second to third prototypes were good. They could just manufacture them because they were all done on the right principles so they threw up factories as soon as our tax credit in California went into effect. They just threw up factories in Denmark and just kept shipping them over. Shipping them over, shipping them over." California slowly filled up with Danish turbines, and even today if you drive over Altamont Pass, due east of the San Francisco Bay Bridge, you can still see some of those first Danish-made, movie-star-owned, S&L-financed turbines chugging away, with their stiff, heavy, inflexible blades. It's been thirty-five years and they still work.

As California's native wind industry was faltering by the end of the 1980s, Denmark's was surging forward, and they colonized the world differently, spreading wind power first to Spain before jumping back over the pond to Texas, Iowa, the Dakotas, and eventually back to California, which in 2015 with 6,018 megawatts of installed wind was the second-largest wind-power-producing state in the country, just after Texas, with 15,635 megawatts, or 10 percent of its in-state generation.

In other words, PURPA worked. There were bumps in the road, great swerving moments at which it seemed that the path toward more efficient, smaller scale, and renewable power generation might be lost completely. But it wasn't. Cogeneration is now a common way to make electricity in the United States, and more important, it has become part of what is commonsensical to us: that excess heat, for example, not be wasted, or that efficiency be a value in and of itself. Likewise, renewables, if not exactly a proven or even cheap technology in the 1980s, had become a way to think about making power. The wind-power boom in California changed Carter's Cardigan Path from a notion to a visible, mappable route leading to new ideas that would follow in the wake of these first, flawed efforts. That even Wall Street accepted that these machines were a way to make money also helped as the moderation of the 1970s slipped over into the resoundingly immoderate 1980s.

Today renewables comprise 13 percent of installed electricity generation in the United States, but that's not news. What is news is that in 2014, 53.3 percent of new generation installed in the United States was either wind or solar, and this percentage is predicted to only grow. Texas is aiming for 75 percent renewable power generation by 2050. The United States as a whole has a more modest goal of 20 percent by 2030, though a recent report from NREL, the National Renewable Energy Laboratory that President Carter helped to establish, holds that we are already capable, with existing technology, of making 80 percent of American power from renewable sources.

What happened in California cemented something of the spirit of the times into national consciousness. There was a better, less harmful, and ideally less costly way to make electricity than with coal, or nuclear or even natural gas. But PURPA was not only important because it changed our ideas about how power might be "better" made, it also opened the door to honest-to-goodness competition in electrical power generation. As bidding auctions between small power providers gradually became the most effective way to integrate new forms of generation, and the companies making it, PURPA helped to prove that bigger wasn't better and that monopoly-governed, vertically integrated, government-regulated megacompanies were far and away not the best way to make and manage American power. Small was not only beautiful but efficient, and, as it has turned out, cost-effective.

The unexpected success of PURPA was a momentous outcome for the culture of electricity in this country. "Passed more on faith than on knowledge of its real potential effects," PURPA helped to destroy the rationale for the regulation of the utilities as natural monopolies. By the early 1990s their monopoly status was indubitably unnatural; it was rather entirely cultural—the remnants of an earlier time when building monopolies was how one did things. What PURPA made clear was that this monopoly structure, at least on the supply side of the system, was a real and proven detriment to the efficient and humane functioning of the business as a whole.

Thus did the 1970s sweep the 1890s from power; the Cardigan Path became the new ideological baseline for thinking as much as making and regulating the electric power industry—though only after it had been adequately proven that smaller, privately owned and managed generation was not only a good way to make electricity but also a good way to make money.

It is often presumed that the Carter-era reforms died a slow death as Reagan-era debauchery and easy wealth expanded to capture the imagination of the nation. We love to read about the excesses of Wall Street as its brokers took a last running gasp at the possibility of outthinking machines. Money flew through fingertips, up noses, between lips and hot, sweaty, very well-paid legs. Overconsumption was back in vogue, too much color (remember neon pink?), way too much hairspray, twenty-four-hour music television, *Miami Vice*, and slippery easy credit. Deregulation was in the air. What hope could a bunch of off-the-grid organic farmers with Ph.D.s and too many pairs of bell-bottoms, like Cashman, have in the face of the brighter, cheaper, gaudier epoch that overflowed the cusp of the 1970s to make the 1980s a decade of big risks in hopes of big returns? The Cardigan

Path, on the surface of things, had been utterly erased by Duran Duran, crack cocaine, and the drive to deregulate and make a profit off everything.

In the electricity business, however, the move to deregulate, which would blossom in earnest in the 1990s, was not entirely counter to the counterculture, for it was regulation that had kept alternative generation out of the grid; it was regulation that had privileged the construction of ever larger, often carelessly polluting power plants through the 1960s and into the 1970s; it was regulation that had made conservation and energy efficiency not matter. If, as one industry critic pointed out, the electricity business is the only one in which you can make a profit by redecorating your office, what did it matter if some people, or even the president himself, were wearing sweaters and turning down their thermostats?

Though it is fashionable to use the term "deregulation" to refer to this epoch in grid history when the utility consensus was first fractured by legislative action, a more accurate term would be something like "reregulation." The National Energy Act of 1978 was a regulatory intervention in the culture of the industry, and it was pushed into efficacy by federal, state, and utility regulators' enthusiasm for the law. Likewise, the next big piece of federal "deregulatory" legislation, the Energy Policy Act of 1992, essentially ushered in a new set of regulatory norms rather than deregulating older ones.

For the utilities, despite the quaking of the ground under their feet, these regulatory shifts from on high also felt a lot like the status quo. They were still being very heavily regulated, just as they had been since the earliest days of the twentieth century, at their own behest. In the intervening years, state and federal political bodies had become used to regulating the utility companies and the utilities had also become very used to being regulated. This did not change with PURPA or with the "deregulatory legislation" that followed in its wake—it felt natural to all parties that any new regulations would oblige, rather than request, utility compliance. This was true at the state and federal level.

PURPA cracked open the utilities' monopsony control of the grid and initiated what would become the end of their "natural" monopoly control over our electricity system as a whole. What it did not change, however, was the prevailing attitude on the part of lawmakers and the utilities that the only way to exert control over the behavior of the latter was through requiring absolute adherence to whatever happened to get legislated. Over time this culture of obligation has had the unfortunate effect of making the utilities even less capable of adapting to the rapidly changing electricity landscape than they had been originally.

When California implemented the first state-level "deregulation" bill aimed squarely at the reform of the utility sector in 1998 several years before the Energy Policy Act became law in the early 2000s they created a debacle, not only because the bill was a very poorly constructed piece of legislation—little more than a remarkable series of loopholes twisted through with some regulatory language—but because it obligated utility and regulatory compliance to its terms.

Or as Fred Pickel, an industry insider during this period, explained to me, deregulation in California failed because it betrayed "an engineering mentality. They tried to set up commercial systems in too much detail and in the process they started up an administrative process in which it takes two years to change a rule."

Sadly for California, in 1998, as the Internet first creeped and then swooped into the national economy, two years was the time scale of dinosaurs. Information was newly electric, and it moved at electricity's own speed, a change of tempo that made loophole exploitation one of the main components of the early Internet economy. The process at the time was called finding "the killer app" (for example, someone figures out that "gas stations in Germany are exempt from the country's rigid early closing laws for most stores. Voilà! German gas stations become virtual shopping malls").

California's "groundbreaking" deregulation bill made it easy, and even fun, to find "the killer app." A number of companies got in the game, none more infamously than Enron. One of the first companies to operationalize online trading of energy futures as well as the exploitation of spot (just-in-time) markets, Enron found some resoundingly killer apps indeed. These included playfully named schemes like Death Star, Ricochet, Fat Boy, and Bigfoot as well more straightforwardly named, equally new, procedures like "megawatt laundering."

Enron was not alone in exploiting the loopholes of California's poorly made deregulation bill; smaller companies in San Francisco were doing the same or very similar kinds of transactions. But Enron's relative size in concert with its broad organizational commitment to making money by whatever means possible lent a monstrousness to its undertakings that other new economy companies could only but aspire to. It has also been useful for the historical record that Enron collapsed, and criminally so, opening their internal workings in 2002 to intense public and legal scrutiny.

Enron's collapse was not the direct result of malfeasance in the manipulation of electricity markets; rather, it was a consequence of their poor management of debt, greed, and risk. Despite the fact that their profitability (or lack thereof) as a company was not primarily rooted in their activities on the ground, these nevertheless have had enduring consequences for the electricity industry. In California these consequences included the near bankruptcy of its two largest investor-owned utilities, the consumption of an $8 billion state budget surplus, and almost nine months in 2000–2001 of uncertain electricity supply as power plants were taken offline for "repairs" or the lines necessary to carry essential current between the northern and southern halves of the state were "overbooked." As rolling blackouts became the rule, many new economy businesses, like Apple and Cisco Systems, as well as other electricity-dependent undertakings such as military bases and prisons, began to think about ways they might detach themselves from grid-provided power. Institutional grid defection in the wake of 2000–2001 became a sign not of radicalism but the inverse: wise organizations engineered ways to use the grid as a backup power system rather than as something upon which they must rely regardless of how poorly it was managed or how sporadically its product was delivered.

Gray Davis, then governor of California, though he would soon be recalled largely because of his failures to contain, or even deal reasonably, with the energy crisis, put it perhaps best as he summed up deregulation in his 2002 State of the State address, saying: "We must face reality: California's deregulation scheme is a colossal and dangerous failure. It has not lowered consumer prices; it has not increased supply. In fact it has resulted in skyrocketing prices, price gouging, and an unreliable supply of electricity. In short, an energy nightmare." On this almost everyone would agree.

It was not, however, "deregulation" per se which caused the debacle, but that the state required the utilities to comply to its poorly crafted legislation while smaller, more flexible, more innovative companies ran rings around them both. This, too, was the result of the culture of PURPA, which presumed the stodginess and unadventurousness of the utilities to be inherent and thus unreformable characteristics of the industry. The utilities could be regulated, wisely or foolishly, but they would not be asked to learn adaptability, flexibility, or creativity themselves; instead they would be told what to do and they would be expected to do it. One result of this is that as the holes PURPA rent in the system have grown larger over the decades, the electrical utilities in their hamstringed attempts to reinvent themselves all too often remain an impediment in the task of reimagining and remaking our grid.

CHAPTER 5

Things Fall Apart

On March 6, 2002, a mere two weeks after a long-delayed inspection, a worker at the Davis-Besse Nuclear Power Station near Toledo, Ohio, found a rust hole the size of a pineapple in the reactor's lid. All that was keeping 87,000 gallons of superheated radioactive water from exploding out of the tank was a thin stainless steel liner. At the time of its discovery the reactor's containment tank had been rusting for years. Already in 2000 a prodigious quantity of that rust and leaking boric acid had been documented, and yet nothing had been done, not by the Nuclear Regulatory Commission (who, it seems, ignored the photographic evidence of severe damage) nor by the utility that once owned but now merely managed the plant and who had repeatedly requested that inspections be delayed or steered inspectors away from its most damaged areas.

In the scrutiny that followed this near to catastrophic eruption of radioactive coolant it was found that Davis-Besse's emergency equipment had been "severely compromised" and could not have been counted on to work properly during a nuclear disaster. It was highly unlikely, had the tank burst, that the operators would have been able to keep the reactor core covered with water and thus cool enough to avoid a nuclear-fueled fire or a meltdown. In 2002,

Davis-Besse became the closest thing to a nuclear disaster in America since the partial meltdown at Three Mile Island in 1979.

Most of us have never heard of this nuclear power plant, one of an aging fleet of a hundred pressurized water reactors in the United States churning out a steady 20 percent of America's power. In 2007 the cooling tower of another reactor, Vermont Yankee, in Vernon, Vermont, actually tipped over. One of the legs holding up this vast vat rusted into unstable flakes of metallic dust while the wooden supports that held it aloft had also rotted through. There was no risk of meltdown in the case of Vermont Yankee's cooling tower collapse, and the thousands of gallons of water that gushed forth were not radioactive; nevertheless, the incident was a spectacular one. This plant, too, was inspected every year and had surveillance cameras focused on the most vulnerable locations collecting endless streams of data for review. The problem is that nuclear plants are huge and complex. It seems that for all their looking, nobody saw the metal rust and nobody saw the wood rot. For the decades it took the elements to eat through the rebar and timber holding up this tank, nobody noticed.

Just as quietly, and just as slowly, the plant was leaking radioactive tritium into the groundwater supply. A leak that despite multiple attempts and even more assurances the plant's managing company never managed to stanch. As a result Vermont Yankee, which produced almost 80 percent of the electricity made in Vermont, was decommissioned in 2014.

There are many stories of this kind. And though heated debates about the safety of nuclear power have been raging in America since these plants were new, there is now little doubt that they are old. Davis-Besse began operating in 1978, Vermont Yankee in 1972. Even some of our younger reactors are causing us problems. Reactors 2 and 3 of Southern California's San Onofre station, both of which came online in the mid-1980s, were shut for good in 2014 after having been beset with maintenance problems in their steam pipes that nobody seemed to be able to figure out how to fix. With this closure Southern California lost 20 percent of its generating capacity.

Even our "young" nuclear power plants are pretty old. And old things break down. They rust through, they develop inexplicable leaks, they function at a fraction of their capacity, and they become increasingly expensive to maintain. A nuclear power plant in this regard is not unlike an aging car. There comes a moment where one has to decide to scuttle the thing rather than continuing to sink money into its repair. The difference, of course, is that upon having donated one's infuriating, money-gobbling, vehicular rust-heap to a local charity one promptly sets about buying another car. Whereas none of America's utilities, with the single exception of the Tennessee Valley Authority (which is to say, the U.S. government) are buying new nuclear power plants. Almost none are buying new coal-burning plants, though year by year the number of these plants in operation is also quietly shrinking. And none are building new large-scale hydroelectric dams. Those heady mid-twentieth-century days of massive, secured investment in big power projects are over. As a result, in 2005, a full fifth of America's power plants were over half a century old and reliant upon technology that was state of the art in the 1950s. More important, as complicated and old as these massive electrical generating factories are, they are but one aging and difficult-to-maintain bit of the older and even more complicated grid that binds them all together. As rightly afraid as we have all learned to be of nuclear disaster, especially after the meltdown of Japan's Fukushima reactors 1 and 3 in 2011, in the United States the fragility of our grid is a more pressing, and more persistent, issue.

This fragility can take many forms. One nuclear power plant may age, leak, break, may be decommissioned, or be taken offline. This is a stress on a grid organized around the presence, rather than the absence, of that source of power. Or viny kudzu may creep with its unerring tenacity to the tops of local utility poles to drape wire and wood and ceramic insulators alike in a thick blanket of green. This, too, is a fragility. Foliage causes shorts and, occasionally, even more dramatic flashovers when our oh so carefully domesticated electricity goes instantly wild again and shoots like a bolt of lightning from a line to an overgrown tree (exploding the latter).

Or ten coal-burning plants may be quietly decommissioned and replaced not by plants of a similar size and make but by smaller generating facilities that run on altogether different kinds of fuel. To continue with the old car metaphor, shutting down San Onofre and building thirty scattered natural gas combustion turbines to replace it is rather more like trading in your 1964 Cadillac Fleetwood for a garage full of jet packs rather than a brand-new Prius. Each of these replacements, while helping to protect us from the pollutants and waste produced by earlier generations of power plants, complicates life for the grid, mostly by introducing variability without storage (chapter 1) and radically distributed, privately owned generation without oversight (chapter 8). The legislative actions put in place to streamline and open up the business end of power production and sale further compromised an infrastructure already weakened by age, decades of incompatible patches, and general inattentiveness.

Despite this, it is hard to be afraid of grid-scale failure. Blackouts, even big ones, rarely last more than two or three days, and they tumble into good times punctuated by fleetingly scary moments and occasional fits of irritation. Money is lost, a few people may die, but for most Americans, blackouts are recalled with real fondness. They are convivial times when strangers come together, ice cream is distributed freely to all, and we eat meat cooked over an open flame. What is weak, what is causal, what is too old or misaligned, is too far from the immediate experience of a blackout to merit attention, especially if accountings are published months (at times years) after the fact. And yet, one would be wise to tremble a little more thoroughly in one's boots when considering the fantastic complexity of the grid that supports us. It is surprisingly easy for little things to go wrong; little things that can become big things with a remarkable speed.

A case in point: On August 14, 2003, eighteen months after Davis-Besse was shut down for repair, the largest blackout in our nation's history, and the third-largest ever in the world, swept across the eastern half of the United States and parts of Canada, blacking out eight states and 50 million people for two days. So thorough and so vast was this cascading blackout that it shows as a visible dip on America's GDP for that year. The blackout, which covered 93,000 square miles, accounted for $6 billion of lost business revenue. If ever it was in doubt, the 2003 blackout proved that at its core America's economy is inexorably, indubitably electric.

Though initial speculation as to cause placed the blame firmly on Canada it was quickly confirmed that this massive outage had been caused by three overgrown trees and a computer bug near Akron, in the territory of the selfsame utility that operated Davis-Besse—Ohio's FirstEnergy.

Not only had FirstEnergy been lax in keeping up with their reactor's integrity, but with deregulation tree trimming, an expensive and thankless task, had slowly fallen by the wayside, as had basic, low-level systems maintenance. And though FirstEnergy is just one of more than 3,300 utilities operating in the United States today, the utility's decisions and failings in the months and days before the blackout—not to mention their behavior once it was in progress—grant us a rare moment of insight into the workings and stumblings of a modern electric company.

The trouble began in earnest in 2001, just as the Energy Policy Act swung at long last out of the contested halls of court into the law of the land. This complex piece of legislation was passed into law in 1992, just fifteen years after PURPA first changed the utility land-scape. After almost a decade of ferocious legal wrangling that took it all the way up to the Supreme Court, the act was implemented in the spring of 2000. Most radically what the act did was oblige the Federal Energy Regulatory Commission (FERC), which governs the grid, to separate electrical generation from electrical distribution. Utilities, obliged since the late 1970s to buy power from small producers at the same price it would have cost them to make it themselves, were, with the act, taken even more thoroughly out of the generation side of the electricity game. The Energy Policy Act mandated absolute competition in the wholesale power market. In many states this also came with an obligation for total, or significant, divestiture in generating stations. From this point forward the main way for the utilities to make money would be by transporting, delivering, and metering electricity—rather than by producing it. That the utilities were shaken up by this organizationally and also fiscally becomes almost immediately evident in FirstEnergy's actions, or rather inactions, in the years immediately following the implementation of this changed policy.

In FirstEnergy's "Annual State of Power Report" for 2001, filed with the Public Utilities Commission of Ohio, or PUCO (a local regulatory body), they had what might be considered a reasonable if not exactly admirable track record of repairs and upkeep. They reported having fixed 75 percent of known "common" problems— like broken wires or poles, leaving 25 percent of these issues for the following year. But in 2002, the year of the near miss at Davis-Besse, they completed only 17 percent of the maintenance and repairs on their To Do list, leaving a backlog of almost eleven thousand items for 2003—the year of the blackout. During the same period, the utility laid off five hundred skilled workers whose jobs were to keep basic infrastructure, such as electrical lines and power plants, in good working order.

More critical to the nation as a whole, however, was a less dramatic-seeming shift in policy: the utility's tree-trimming schedule was changed from every three years to every five. Given that a utility's compliance with their own internal policies regarding tree trimming is voluntary (which is to say, they are encouraged to stick to it but there is no oversight or sanction structure in place if they don't), even checking and cutting back trees every three years can mean that particular points on the system, especially those on private property, can go much longer between prunings. The tree that started the 2003 blackout was almost fifty feet tall. And while there was at the time no national standard for how tall trees near high-voltage lines might be allowed to grow, it is generally agreed that fifty feet is about a decade more growth than is acceptable.

FirstEnergy was hardly the only utility to have cut back on foliage maintenance; tree-trimming budgets, while not gigantic, do form an easy source of revenue for utilities tightening their belts or, as was the case of PG&E in California in 1994, expanding executive pay.

It is rare for cutbacks and cash transfers of this kind to rise to the level of criminal negligence, making it a relatively safe pool of money to siphon off for more pressing purposes. Though we do not know how much FirstEnergy saved by extending its trimming schedule by two years and then not actually sticking to it, we do know that PG&E managed to transfer about 80 million ratepayer dollars into directors' and shareholders' pocketbooks using the same kinds of minor shifts in scheduling. Unfortunately for sparsely populated Nevada County, California, they also caused a massive, fast-burning fire that destroyed more than five hundred acres—a fire that started when a treetop brushed up against a power line and burst into flames. This offending tree was one of 767 violations in that county alone, all of which PG&E knew about and ignored, while simultaneously reducing tree-trimming crews from three to two men, shifting their cutting schedule from three to five years, and lobbying the California legislature to change the recommended clearance between the treetops and power lines from four feet to just six inches. The Trauner fire of 2004, much like the East Coast blackout of 2003, was caused by a tree that was at least a decade behind on its trimming schedule.

Had tragedy not struck in the Ohio and California cases, no one would ever have been the wiser: the trees grow a little bit, the budgets work out more to the liking of one party or another, and business continues as usual. Many U.S. utilities "adjust" their tree-trimming and their fiscal budgets in similar ways. This despite the fact that there is something utilities know and that their customers (including most of Congress) remain ignorant of—in the United States today, as in 2003 (and as in 1994), the greatest threat to the security and reliability of our electrical infrastructure is foliage. Trees most especially, though kudzu and its ilk are troublesome creepers in their own right.

Politicians may talk a lot, and utility managers may worry a lot, about how terrorists might hack into, shoot up, or bomb various bits of our grid in order to bring the United States to her knees. And yet the trees constitute a far more significant threat to the security and reliability of our national electric infrastructure. It is every utility's responsibility to keep the plants away from the electric lines; this is as true on public lands as on privately owned farmsteads.

Every utility knows the stakes, and yet trees still cause most of our power outages. During storms, trees and branches falling on lines do most of the damage. Even in calm weather trees regularly tangle into household distribution lines, shorting them out or ripping them down. These are small-scale issues compared to what trees do to the high-voltage lines that run from power plants to periurban substations. These long-distance lines are the arterial system of our grid. Bring them down, short them out, or cause them to trip offline by asking them to carry more electrical current than they are able and the system as a whole is thrown into an intense state of risk. This is what happened in 2003 in Ohio. Not just one but three separate trees came into mortal contact with a line.

The first, a 345-kilovolt line, "arced" in the suburban Ohio subdivision of Walton Hills, shooting a bolt of energy much like lightning out at the closest thing around—in this case, the tip of an overgrown tree. This bolt of artificial lightning was accompanied by two thunderous bangs and, shortly thereafter, the explosion of the Muha family's dishwasher.

Adam Muha, then eighteen years old, lived with his parents just under the now nonoperational line and next to the offending tree. He'd just gotten home from school when the line arced. As surprised as he was by the noise, he was even more stunned that most of his family's home appliances and two of its outlets began billowing smoke. Going outside to investigate, Muha was met by a tree-trimming team that by chance had been working across the street (investigations have proved that their presence on the scene was incidental, not causal). One of the men ran toward Muha when he saw the boy standing somewhat confusedly in his driveway. "Get of here!" he yelled. "If that line comes down your car will melt, our truck will melt, and we will melt." Muha obliged.

There is nothing unusual in the story so far: a hot day, an overtall tree, a sagging wire, a flashover, which, while bad for both the tree and the line, was hardly a harbinger of the violent voltage swings—and resultant blackout—to come. This kind of death of a wire happens all the time in the United States. A tree, a wire, a bang, a short, followed by the automatic shift of the dead wire's "load"—this is the technical term for the electricity it is asked to carry—to another, duplicate wire, heading in the same direction. This kind of redundancy on the system, which has been intentionally increased since 1968 after the Northeast's first big blackout, is today de rigueur. Even most rural systems have some duplication of duties on their higher voltage lines. This doesn't, however, mean that there is an indefinite number of available lines; there are more than we need for everyday power transmission and distribution, but this doesn't always add up to enough for exceptional events.

All of this redundancy absorbed the loss of that first line in northern Ohio. Its failure was mostly a nuisance. The problem came later. And when speaking of a product that moves at nearly the speed of light, later came on very quickly indeed.

A second wire also sagged into a tree, waiting there, too tall, its branches stretched skyward to catch every ray of sun and channel every drop of rain downward to its thirsty summer roots. This wire, like the first, was warm and sagging low with the heat of the day; it was, like all conductors, apt to stretch longer in hot weather. This expansion is a molecular property of metals that is exacerbated, but not caused, by the presence of an electric current. All the lines in Ohio, in the Midwest, in the East, were hanging low that day as they were on any day in any August. Had the tree not been there, or had it been trimmed down, both it and the line would have survived the limpid afternoon, rather than a *bang!* and a fizzle and the smell of charred wood. And a second line was out. This one, the Chamberlin-Harding line, ran just south of Cleveland. Its load, too, was transferred to yet another set of wires, headed yet again in the same direction. And thus did the cascade begin. A load easily borne by one wire or two proved too much for the third duplicate.

Every line has a rating, a voltage, a "quantity" of electricity that it can safely conduct from one place to another. Extra High Voltage lines, the ones threaded across open prairies and high mountain passes borne up by huge steel towers, carry from 275 kilovolts (kV) to 765 kV, while those that link the corner pole to the side of your house have a low voltage rating, usually around 50 kV. Lines are actually pretty remarkable technology even if they all look the same from ground level; its the shape of the poles that allow average folks to differentiate one voltage rating from the next, but up there the lines that are strung between these poles are a complex assortment of alloys and weaves and nested arrangements of metals, each designed to carry its assigned current safely from the point of production to the point of use.

The first line that met that first tree out there in Walden Hills, Ohio, was a 345-kV line. This is a midsized long-distance transmission line carrying AC power, as was the second, as was the third (this outage, too, was caused by a tree—you begin, I hope, to see the trouble with trees). It is here, about an hour into what was not yet but was fast becoming a blackout, that automation began to get seriously involved. Within ten minutes of the third deadly tango between line and tree, a fourth 345-kV line was lost—this one because it was being asked to carry more current than it could safely transport—followed almost immediately by fifteen 138-kV lines shutting themselves off automatically. This is a self-preservation technique designed into electrical conductors ("conductor" is the technical term for an electric transmission line) when giant waves of unstable voltage threaten to fry, melt, or otherwise disable them. At this point, 3:41 P.M., about thirty minutes before the actual blackout started, trees ceased being the problem. Line failures were fast becoming commonplace because circuit breakers had begun to trip.

Much like the fuse box in your home, transmission and distribution lines are armed with a delicate, and fairly simple, analog system (that means no computers are involved) for keeping lines in good working order. If, for example, the electrical current a line is asked to carry exceeds its rating, its circuit breaker trips, the line is taken "offline," and its load is transferred to a duplicate line. Unfortunately, by about 3:41 P.M. there weren't many lines left, duplicate or otherwise, and within a few minutes of the circuit breakers' snapping open, cutting off current from at-risk conductors, an interesting thing began to happen to the grid: it became unbalanced.

Like a madman or -woman suffering from some sort of inner-ear condition, the lack of available transmission caused a lopsided high-speed wobble of the queerest sort. It seems that there was way too much current on what little transmission capacity remained to carry it—a condition called overcurrent that causes any low-voltage area, in this case almost all of Ohio, to work like a giant suck on any and all the electricity elsewhere in the network. To understand this we have to pause a moment to talk about electricity, in all its weirdness, again.

As we know from chapter 2, electricity does not move by human logic—it does not, for example, take the shortest path between two points. Nor does it move by water's logic, though there are certain similarities—it does not, for example, "flow downhill" or "puddle" on even terrain. Nicely, too, it won't drip out of an "open" outlet. Rather, electricity will take all possible routes available to it simultaneously with a preference for that with the least resistance even when that is not the shortest or most rational (to our minds) way to move from point A to point B. The routes we'd prefer electricity to travel are built from highly conductive materials and those we'd rather it avoided are made of materials with low conductivity. Air is not very conductive; this is why the electricity in your outlet doesn't seep out. Metal, as we know, is a lovely conductor, hence the long-standing reliance on transmission lines as a very good way to corral and direct electric current from where it is made to where it is used (and incidentally why we are trained from an early age not to stick a fork in an outlet).

To travel around Lake Erie, an electric current made just outside Toronto will simultaneously take the shortest path to New York City (about five hundred miles) and it will take the longest path, winding its way through Michigan, down to Kentucky, over to Virginia, back up through Maryland and New Jersey to finally arrive in New York. It will also take every single other path, with a decided preference for low resistance routes.

So long as resistance is equal on all paths the electricity that takes a longer, more confused or circuitous route will arrive at the same instant as the one that took the shorter path. Distance is irrelevant to it, only resistance matters.

The reason a bulb illuminates when you flip on a light switch is that you have just lowered the resistance on the circuit, from total (when the switch is off) to near zero (when the switch and lamp are on).

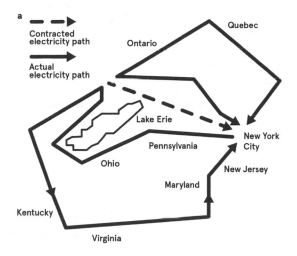

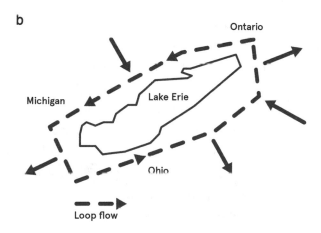

FIG 4 Electric power does not travel just by the shortest route from source to sink (*top*), but also by parallel flow paths through other parts of the system. Where the network jogs around large geographical obstacles, such as the Rocky Mountains in the West or the Great Lakes in the East, loop flows appear that can drive as much as 1 GW of power in a circle (*bottom*), taking up transmission-line capacity without delivering power to consumers. (*Reproduced with permission from "What's wrong with the electric grid?" Eric J. Lerner. Published online with* Physics Today *on August 14, 2014. Copyright 2014, American Institute of Physics*)

There is no need to "communicate" with electricity in any other way—our grid, indeed the whole of our world, might be thought of as a symphony of varying resistances, each beckoning or shunning electric current in its own way.

This beckoning, called a "sink," says to all the current on the grid: "look here, this is the easy way." To get electricity to where we need it we make little sinks, like flipping on a light switch, and big ones like firing up a paper mill. The real-world ramifications of this simple physical law are flabbergasting in a system as complex as our electric grid. As one physicist explained, "any change in generation or transmission at any point in the system will change loads on generators and transmission lines at every other point—often in ways that are not anticipated or controlled."

On August 14, a generating plant in Eastlake, Ohio, was down; so was Davis-Besse. This meant that electricity was already moving through the grid differently from how it would have been had these two generators been in good working order. That the grid is so large, so interwoven, with so many hundreds of generating plants, so many thousands of miles of lines, so many utilities and other power-managing entities—all this means that it simply can't be modeled in real time. The best we can do is monitor, computerize, and react to it as quickly as possible when the situation changes.

But FirstEnergy didn't react. One line went down, a second, a third, a fourth, a fifth, and then simultaneously fifteen more. Twenty high- and medium-voltage lines down and no word is coming out of FirstEnergy's control room. They aren't doing anything. They aren't calling anybody. They are hardly reacting when others call them. All because of a problem of a different sort. They have a software bug—a tiny, fairly innocuous, incompatible line of code that slowed, perhaps even halted, the refresh rate on their computer screens while also silencing their alarms. They didn't know they'd lost all these lines; they were blind to the blackout that was coming. And because they didn't see it, neither did anyone else.

What FirstEnergy couldn't see electricity itself was already attempting to right, but in its own way, which in this instance proved to be most unhelpful to the grid and its human operators. Ohio, thirty minutes before the East Coast blackout began, had transformed into an enormous sink. From electricity's point of view it looked as if there was not enough available current to meet demand. This was, in a way, true. Enough lines were down that the current that was stranded on the system just kept overloading and incapacitating the lines available to it; it couldn't get to where it was needed. The sink remained despite the fact that too much current, rather than too little, was causing it. What the electricity being produced by every power plant around Lake Erie and up into Michigan and over into New York "saw" was a path of least resistance leading right into the heart of Ohio. Ohio looked like a sink, it acted like a sink, but it was not a sink. This confused the current, which began to surge first in one direction and then in another, rather like tsunami-sized waves sloshing around in a bathtub.

So New York State, for example, instead of having been duly informed that something was amiss in the Midwest in the usual ways (a phone call from someone at FirstEnergy would have been best), got their first hint of the blackout to come when 800 megawatts surged suddenly westward—effectively being sucked out of their system toward Ohio—and then abruptly reversed direction and shot back into New York when the sink it was reacting to proved false. This is the equivalent shift in voltage that they would have seen if someone had flipped the Off switch of a nuclear power plant the size of Davis-Besse and then, having realized their error, flipped it right back on again. Such swings are not good for the continued longevity of our already enfeebled infrastructure.

To protect themselves and their machines, operators in Albany started shutting down generating plants and isolating transmission lines across the Northeast. As in New York, so also in Pennsylvania, the western half of Michigan, Ontario, all trying to "island" themselves from Ohio and its great energy sink. This process of islanding, or decoupling one's local grid from the system as a whole, is not easy given the current organization and size of our interconnections. And though it was a wise protective measure (in 2003 islanding kept a great many power plants and power lines in New York State from being destroyed by massive swings in voltage), it also caused the blackout in that state. As soon as New York was islanded, it was producing far more power locally than it needed, and since it's a much slower process to turn down a coal-burning power plant than it is to island a bit of the grid, their lines, like Ohio's and Michigan's and Ontario's, began to automatically isolate themselves from the dangerous quantities of electricity they were suddenly being asked to carry to nowhere in particular. So much power was moving around so fast from one imbalance to another that American Electric Power (FirstEnergy's neighboring utility to the west) reported that one line was carrying 332 megavolt-amperes (MVA) despite being rated for only 197 MVA.

After having put in several panicked phone calls to FirstEnergy during the hour and a half before the blackout smashed into the Eastern Seaboard, operators at American Electric Power also chose to island themselves—closing their entire network to the erratic surges of current they saw coming toward them. Because they were closest to FirstEnergy and because they were among the first to understand that FirstEnergy had no cognizance of the situation unfolding in their midst, the western edge of the blackout was precisely the line between FirstEnergy's operating area and that governed by American Electric Power.

Nor were befuddled engineers from American Electric the only ones to have called FirstEnergy in the hours leading up to the blackout. They got a lot of phone calls that day before things went critical and went black.

At three P.M., almost exactly one hour after that first line in Walden Hills arced, alarms began to clatter and complain in the municipal service district headquarters in tiny Wadsworth, Ohio. Gene Post, the electric superintendent for the borough, called FirstEnergy to ask what was up. "They said they were in the dark," Post said. Not literally, of course, not yet. He meant that they didn't know what was going on. Worse, they didn't seem to know that anything *was* going on.

"They couldn't give us any information." Post was befuddled. The swings on the grid were remarkable. Unusual. Dangerous. Certainly FirstEnergy ought to have noticed! If anyone had information it was them, the guys who owned and managed the lines. These are exactly the circumstances, Post said, "when we expect someone to know what is going on!"

When Steve Dupee, the director of Oberlin Municipal Light and Power, called FirstEnergy about thirty minutes later to ask why Oberlin was experiencing such extremely low voltage, "the guy told us he didn't know what was wrong, because his computer was down."

Two minutes later, FirstEnergy got an even more worrisome call, this one from the organization charged with monitoring and regulating all the power in the Midwest as it moves across state lines—the Midwest Independent Transmission System, or MISO. When MISO operator Don Hunter called in just after 3:30 P.M. he got a befuddled Jerry Snickey, an engineer with FirstEnergy.

Hunter started with the obvious losses: "I think what happened here is that you guys lost the . . . South Canton to Star, and also the Hanna-Juniper line . . . at the same time roughly."

To which Snickey replied: "You show the Hanna Juniper line out also?! We have no clue . . . Our computer is giving us fits. We don't even know the status of the stuff around us."

Hunter, frustrated: "I called you guys like ten minutes ago, and I thought you were figuring out what was going on there!"

"Well we're trying to!" sniped Snickey, also clearly irritated: "Our computer . . . it's not happy . . . It's not cooperating."

This was at 3:32.

By 3:45, when a foreman of Cuyahoga Falls's electric system called because a low-voltage alarm had gone off on his system, FirstEnergy clearly knew something had gone terribly awry. Basically, he was told, "There is a problem, now hang up the phone."

FirstEnergy had gotten the message. But an hour and thirty minutes between the loss of the first line and that moment of awareness was an hour and thirty minutes too long. And, worse, for sixty minutes or so—and this is gossip, mind you, not fact—system monitors seemed not to have noticed that something was wrong with the computers. Critical information was not being highlighted or even displayed, the refresh rate had slowed dramatically; it may even have stopped altogether. In other words, there was enough *not* going on with those machines that somebody with a careful eye or a bit of concerted attention should have been able to see that they were behaving differently than on a typical day.

To their credit, the control rooms of power companies are full of screens, and these are full of data, charts and colors and rows of text, and control room operators are highly dependent on alarm systems that are either audible or flash on their screen, calling attention to critical changes in the flow of information. Without these alarms it's questionable whether any normal human being is capable of noticing critical shifts in data patterns or laggardly mechanical behavior. And the alarms that day in FirstEnergy's system didn't work. Neither the onscreen alarms nor the speaker-blasted ones. There was no reason for anyone to have paid attention to the fact that their screens weren't refreshing. Those screens said everything was A-OK. Before the stream of phone calls began, it seemed entirely reasonable to believe that this was the case. August 14, well into the chaos that would become the blackout, was for FirstEnergy's control room operators just another day at the office.

All the while the bug was there. Not a bit of malware, not a virus, not a worm. Just a glitch, more of a programming error than anything else. In effect what this bug, called XA/21, did was cause a machine to respond with a busy signal when multiple systems tried to access it simultaneously rather than prioritizing these requests and then taking each of the "calls" in turn. As more and more data points were rebuffed, they started to stack up rather than being logged and then deleted. Like silt in a water purification system or cholesterol in your artery, all these tiny bits of retained information started to stanch the free flow of all information, slowing everything down, way down, and eventually crashing the main server. All the accumulated unprocessed events were then transferred to the backup server, which was no better equipped than its predecessor to handle this vast backlog and so it, too, failed. At which point, an hour or so after the first line failure in Walden Hills, right about the time Hunter from MISO was calling in because the situation had become very dire indeed, the control room operators at FirstEnergy finally realized something was terribly wrong.

Though there are many differences between the electric grid and the microelectronic processing equipment we use to monitor and run most of modern life—including the grid itself—there are a couple of commonalities worth mentioning. First, both rely upon the same physics. Information these days is also electricity. Second, as the case of the XA/21 bug makes very clear, cascading outages can and do happen on both information systems (computers mostly) and electrical systems if relatively minor errors are not acted upon with all due haste.

In the case of the 2003 blackout the error on the grid took the form of overgrown trees and the error on the computers took the form of a line of code that disallowed simultaneous incoming data reports. Each error had small but significant effects on the actual physical infrastructure they came into contact with, while the way in which this physical infrastructure was designed—its logic—allowed for stochastic propagation of negative effects. Which is to say: the cascade and then the blackout.

One electrical line and one tree became three different lines and three different trees became fifteen lines overloaded and automatically tripped; pure mechanical self-preservation becomes great peaks and valleys in voltage in Ohio and Michigan which becomes vast waves of voltages crashing across states all over the Eastern Seaboard, smashing into generating stations, substations, high-voltage wires and low-voltage lines, disabling them all. All of which becomes the oft-quoted $60,000 per hour per business of lost revenues and an upsurge in blackout-conceived babies nine months later.

A single line of code, in a program comprised of a million such lines, a fluke lasting little more than a microsecond, causes an alarm-event application to go into a loop. It just keeps spinning, spitting that bit of data back at the server, getting a busy single, spinning and spitting, as the data not making it through mushroom from bits to bytes. This spinning manifests itself in slow computers, frozen screens displaying perfectly acceptable conditions as the grid outside the window is degrading ever more quickly. It becomes alarms that don't sound, which become systems operators that look at their screens and tell worried callers that everything is fine. Until there are enough worried callers, until the data glut has gotten so severe that the servers have crashed. Until it is too late to do anything about it. Within twenty minutes of that last phone call, FirstEnergy's own command and control center had blacked out.

It was too late. The blackout had begun.

This was not the first major blackout in the East, nor will it be the last. In 2003 no one in particular was to blame. Nobody noticed the tree as it grew, nobody noticed the bug as it slowly gummed up the works. Like rust and rot, like tiny leaks and hairline cracks, like age itself, the tree and bug were too minor and too quiet to catch anyone's eye. And yet there they stand at the beginning of the cascade, singular monuments to all the smallnesses that can add up, with time and opportunity, to total systems collapse.

One might be given to think that this blackout might have been prevented if somebody had just noticed as things slowly went awry—if in 2002 all of FirstEnergy's "known common problems" had been dealt with rather than merely 17 percent of them, if the trees had been clipped, if a bright young eye had seen the static in the screen. But what most students of industrial accidents recognize is that perfect knowledge of complex systems is not actually the best way to make these systems safe and reliable. In part because perfect real-time knowledge is extremely difficult to come by, not only for the grid but for other dangerous yet necessary elements of modern life—like airplanes and nuclear power plants. One can just never be sure that every single bit of necessary information is being accurately tracked (and God knows what havoc those missing bits are wreaking while the presumed to-be-known bits chug along their orderly way). Even if we could eliminate all the "unknown unknowns" (to borrow a phrase from Donald Rumsfeld) from systems engineering—and we can't—there would still be a serious problem to contend with, and that is how even closely monitored elements interact with each other in real time. And of course humans, who are always also component parts of these systems, rarely function as predictably as even the shoddiest of mechanical elements.

Rather than attempting the impossible feat of perfect control grounded in perfect information, complex industrial undertakings have for decades been veering toward another model for avoiding serious disaster. This would also seem to be the right approach for the grid, as its premise is that imperfect knowledge should not impede safe, steady functioning. The so-called Swiss Cheese Model of Industrial Accidents assumes glitches all over the place, tiny little failures or unpredicted oddities as a normal side effect of complexity. Rather than trying to "know and control" systems designers attempt to build, manage, and regulate complexity in such a way that small things are significantly impeded on their path to becoming catastrophically massive things. Three trees and a bug shouldn't black out half the country.

Or, to put it differently, one doesn't try to eliminate the holes in the Swiss cheese—by compacting all cheese into cheddar, for example—but rather to keep the holes in the cheese from lining up, from becoming one big hole that runs through the entirety of the loaf. The holes will be there (trees will be too tall, budgets will be tampered with, regulators will be put off, computer bugs will worm their way in), but they won't snowball into total systems collapse the way they did in August 2003, and the way they very nearly did at Davis-Besse a year earlier.

The problem with the Swiss cheese model, which is otherwise remarkably robust (we have it to thank for the past thirty years in which flying has been consistently safer than driving; even frequent fliers are more likely to be killed in a lawn mower accident than an airplane crash), is that the grid, like any complex mechanical system, is not just a machine but also the regulatory, business, cultural, and natural environments within which this machine functions. The grid is the physics, mechanics, engineering, construction, management, upkeep, and use internal to itself. It is also the storms, earthquakes, laws, hatreds (and other personal opinions), and profit motives that surround the mechanism, that change over time, and that can differ drastically between one state, one city, one climactic zone and the next. In the case of the grid these "external to the cheese" circumstances are not necessarily conducive to the machine's capacity to function well and strongly.

If the 2003 blackout was traced back to particular failings on the system—the holes in the cheese of our grid—and if a number of these failings have been since corrected, this tracing out of proximate causes to a single, massive systems failure doesn't necessarily direct our gaze in the right direction, or set our minds to the right problems.

No amount of precision in the chain of events that led to the 2003 blackout, much detailed and investigated after the fact, will lead one to the whole story of why it happened where and when it did.

For that we have to look beyond the machine into the vagaries of American political and economic life. For what is perhaps most remarkable was that it was a disaster almost entirely devoid of criminal acts, but which was, nevertheless, thoroughly man-made. This differentiates 2003 from more catastrophic events like Hurricane Katrina and Superstorm Sandy, which were natural disasters first, if rendered more extreme in their effects by an immensity of human error, and from outages like those in California in 2000–2001, which were criminally manufactured by Enron and other energy trading companies operating in that state. In Ohio in 2003 the outage was man-made from the first fault, trip, and short to the last, yet there was no causal malfeasance associated with FirstEnergy's many failings the long August afternoon the East went dark.

Shortly after the release of the first government report on the blackout, which apportioned blame almost exclusively to FirstEnergy, the vice president of energy delivery for the company, Chuck Jones, did as one might expect—he "vigorously defended the company against allegations that it was the primary cause of the blackout."

It would seem a weak, indeed indefensible position, like preparing for a critical battle by stationing all your soldiers in the bottom of a valley. Jones's repartee sounds so pat, so predictable, that it blurs into the white noise of self-defense that corporations often make when claiming innocence. It's not our fault, Jones said. "Too much electricity flowing over an antiquated grid was really to blame." It's the grid, he said, that's the problem.

Despite how habitual it has become for most Americans to disregard claims of innocence made by corporate spokesmen, in this case Chuck Jones actually has a point. Whether or not FirstEnergy should have let one—not to mention three—trees grow over forty-five feet near power lines, whether or not their computers were buggy and their control room operators too invested in coffee and distraction to notice that their screens were not refreshing, whether or not FirstEnergy is to blame for all the proximate causes of the outage, and they are, Chuck Jones still has a point.

The problem is the grid. And not just because the technology is old, not just because it's insanely complicated, and confusingly managed, and not just because when something goes wrong our number one most effective communication tool is still a telephone.

This is not why we should give credence to Jones's words.

In order to understand why the grid has become so much less stable since the early 2000s (and it has), it is important to return again to a more careful consideration of the aftereffects of the Energy Policy Act and accompanying Order 888. Much like the 1996 deregulation bill in California (that made Enron momentarily very rich and that state by equal measure poor), the Energy Policy Act did not separate generation from transmission and distribution just for shits and giggles. It did so for a reason, and that reason was energy trading. The act turned electricity into a commodity—a thing like any other. While the grid, electricity's infrastructure, was reconceptualized in law to something rather like a box or shipping container. Conceptually if not actually, electricity now is made and sold in units to whoever pays the best price and then shipped to them by means of the grid.

At the time, Jack Casazza, an executive at PSE&G in New Jersey said that the lack of recognition of "the single machine characteristics of the electric power network" enshrined into law with these policies is rather like "having every player in an orchestra use their own tunes."

Within a month of the Energy Policy Act becoming the law of the land electricity trading had shot through the roof. There was energy trading previous to the act but it had been a minor part of a utility's operations and often looked a lot more like horse trading than modern arbitrage. A trader at a utility in a pinch might call a paper mill and say: Hey, if you guys turn your mill off for twelve hours, we'll get the guys at the chloramine plant, who need a little more power right now, to give you 40 tons of chlorine bleach for free.

After the act, the situation could not have been more different, foremost because the price for electricity was, for the first time ever, being set by traders tying to anticipate demand in certain locations, under certain conditions, and at certain times of the day. They would buy the cheapest electricity they could find and "wheel" (transport) it to be sold in the most lucrative market. Electricity was now and newly both a real commodity—tradeable, transportable, and profitable—and a real, hedgible financial entity with futures markets and derivatives all its own. Electrons, from the point of view of the market, have never looked so much like pork bellies or pig iron.

The idea behind the act was twofold. First, it would liberalize and thus also reform (by means of the market) electricity production. This has indeed happened. There are many more kinds of fuel feeding the power plants that supply the grid today than ever before, and much of the innovation we see in power production—from the rampant adoption of rooftop solar to the new popularity of natural gas fracking—is the direct result of competition in this sector. And second, it was to make a unit of electricity into a tradable commodity with a price set by relations between supply and demand rather than by a utility's regulatory body or by convention. This is literally what deregulation means when applied to the electric industry. In addition to the loss of regulatory support for both monopsony and monopoly powers, the prices utilities pay for and the price they charge for electricity have become far more volatile. Market forces now matter to the utilities. Regulators as well as state and city governments still have a whole lot of say over what a utility can and cannot do, what they must do, and how much they can charge for any of it, but this "a lot" is a whole lot less than it used to be. The introduction of market forces to grid management has, after some initial bumps in the road, also had a profound effect both on how much electricity we as a nation use (less) and on the way that electricity moves through the grid (farther).

Both the possibility of trading electricity and of making it by whatever means strikes one's fancy has inexorably changed the work done by our grid without changing its structure very much. How this electricity was moved from cheap "there" to expensive "here" almost immediately began to prove a problem. As was mentioned earlier, all electric lines, called conductors, are rated according to the maximum voltage they can safely carry. Their structure reflects this rating, as do the regulations that come with it: the minimum distance of the lowest conductor from the ground, the maximum height of nearby trees, and the kinds and strengths of the towers these lines are strung on.

More poetically and of equal technical relevance is that different kinds of electric lines are not identified by long strings of numbers and letters but rather are named after birds. So, for example, a 765 kilovolt (kV) partridge line that runs one hundred miles has a maximum carrying capacity of 3.8 gigawatts (GW)—or 3.8 billion watts, roughly the electricity needed to light 10 million 100-watt bulbs or all the homes in Columbus, Ohio. Now take that same partridge conductor, stretch it to four hundred miles, and it can carry only 200 GW, about half as much power. Distance reduces the carrying capacity of lines, whereas long-distance wheeling increases the amount of electricity transported over precisely these same lines. The Energy Policy Act did nothing to upgrade the lines. The result was a simple one. The free trade of electricity meant that there was too much electricity traveling too far. Lines were getting overburdened, heating up, sagging, shorting out, arcing, and filling with harmonic resonances. All of which are both very bad for reliability and wasteful, and thus bad for efforts at conservation.

In May 2000, a mere two months after the order that utilities "wheel" electric power rather than make it went into force, the number of

Transmission Loading Relief Procedures on the East Coast's grid was six times what it had been a year earlier. These requests for help managing the load of particular lines are one very good way of measuring stress on the grid. Just before the act was made law, the number of these requests on the Eastern Interconnection was at right around ten per month. Six months later, in August 2000, there were almost 175 such requests, and by mid-2003, right around the time of the blackout, more than 250 such requests were coming in per month.

Problems of this kind with line load could lead to unusual flash-overs in places where one would not expect to see them, places like suburban Walden Hills. There was just too much electricity moving too far on the grid for it to behave in the same, somewhat predictable ways it had when power production and transmission were both more local phenomena.

"Predictability" is an awkward term when speaking of electricity, because strictly speaking it isn't. Previously, predictability had been premised upon an algorithmic sense of user behavior and existing patterns of generation, transmission, and consumption that follow from population density, the season, the workday, and other fairly stable indicators of when and how much power might be used. The utilities made their generation and investment decisions based on twelve data points per year per customer (the monthly meter read-ings). Long-distance wheeling and the unlinking of generation from local consumption habits has meant that a lot more data is now needed to manage the electricity moving on the system at any given moment. Utilities, however, are not masters of data. Nor is it exactly the case that someone in a control room somewhere can use this data to "see" the grid in all its real-time complexity. What is managed is a guessed-at set of demands, and of supplies, and of line carrying capacity. When these guesses go wrong, that's when the alarm bells should start to ring.

Indeed, this "guessed-at" way of insuring grid stability was largely how Enron managed to manipulate California's grid so profitably. For example, one way, among the many, that they made money was to purchase transmission rights on essential power lines, most significantly a thousand of the 1,600 possible megawatts on California's path 26, one of two conductors linking the northern and southern halves of the state. They could then fictitiously "clog" this route. They did this not by adding any actual electricity into the grid, but by simply saying there was no more room available on these critical paths and then getting the state to pay them to "free up" transmission capacity that had in fact been there all along.

Enron's market-manipulating machinations to the side, almost all contemporary electricity traders engage in forms of electricity arbitrage (buying cheap here, selling dear there) made possible by the critical combination of the Internet and the Energy Policy Act. This way of making money on power may have been first exploited by Enron—in the late 1990s, 25 percent of all energy trading in the United States was being conducted via EnronOnline—but it also brought competition to the market in a way never before seen. The unfortunate side effect of this was to make the flow of information less free as the flow of electricity has become simultaneously more complex.

Utilities used to talk to one another. No one ever made money off of anybody else's customer base and so grid instability in any one region could serve as an object lesson for utilities dealing with the same issues in their own locales. Since 2000, however, not only is there far more information pouring into utility databases—a product of the widespread introduction of computerization at many points in the system, of which digital "smart" meters are the most infamous— but that information is treated as proprietary. Electricity trading, in this way, fails to be laissez-faire capitalism because nobody quite has access to the information necessary to make real-time decisions about how to manage their systems, and thus also their capital. Not even, oddly, the utilities themselves, since they now have so much information to contend with that most of it sits unprocessed in giant servers called "historians." Big data has become just another modern way to use up electricity.

Beginning is the early 1900s utilities made decisions about how electricity was generated and how it moved on their slice of the grid— information they freely shared among themselves. Now they neither make these decisions nor share information about them. Historically, utilities made money when people used electricity; the more we used the more money they made. Now they don't. Today's utilities make money by transporting power and by trading it as a commodity. While they are still charged with keeping America's power supply reliable they have a real incentive to sell electricity to whomever will pay the most for it wherever they may be. Long-distance wheeling is to their benefit; it is to the plant owners' benefit; it is to the energy traders' benefit; in theory, at least, it is also to our benefit. In fact, the only thing that really suffers from this arrangement is the grid.

This wouldn't be the case if we put a lot of money into upgrading the grid—making it the technological equal of the new regime of power that governs it. The problem is that utility companies' profit margins have been shrinking, not just since the passage of the act into law, but since the late 1960s. Then, overinvestment in nuclear power coupled with the nascent conservation movement led to the surprising trend—for Americans—of decreased consumption. By the late 1970s, big capital investments in electrical infrastructure had slowed to a trickle, where it remained until the act made it startlingly lucrative to start buying or building new generation facilities again. Many of these brand-new power plants are much smaller, more dispersed, and more variable than anything the grid has ever seen. Natural gas combustion plants, solar panels on rooftops, wind farms in the Dakotas, all are forms of privately owned power that further complicate the grid while, finally, breathing new life into the electricity production game. But these new constructions are as often owned by investment firms, like.

If we follow the flow of modern money, a new terrain of investment and of privation emerges into view. And where money doesn't go, where people don't want to spend, starts to give us a good idea of which bits of our grid are given to falling apart. This is the landscape of collapse—imminent and actual.

Money doesn't go to the upkeep of the fleet of old, lumbering power plants trudging toward retirement that nevertheless still form the backbone of America's electrical generation facilities. In the case of nuclear this lack of desire to make significant investments in upkeep is the most worrying because the results of plant failure can be the most catastrophic. This isn't to say there is no capital invested in this domain—far from it—it's just not where the utilities *want* to spend their diminishing returns. FirstEnergy is currently spending another $600 million to upgrade Davis-Besse's steam generators, this after having spent an equal amount in 2004 to replace the perforated reactor head. Given the cost, one can easily imagine why they diverted inspectors' gazes elsewhere and requested, on multiple occasions, to have inspections delayed or canceled as rust slowly ate into the nuclear containment vessel; for this they were fined $5 million by the Nuclear Regulatory Commission, the largest fine the NRC has ever levied, and they paid another $28 million to the U.S. Department of Justice for deliberate obfuscation of the facts.

In the decades preceding the 2003 blackout money didn't flow toward the upkeep and new construction of long-distance transmission lines. It is incontrovertible that our wires were underprepared for long-distance wheeling, just as they were unprepared for more predictable things like population growth, the veritable explosion of pluginable devices that has come with the revolution in telecommunications technology, and our seemingly insatiable hunger for air conditioners. Not to mention the worrisome possibility that our cars, too, might be electric one day soon.

Since 2003, however, the transmission portion of our system has gotten a lot of attention and more than its fair share of the cash. Hundreds of thousands of microsensors have been placed on lines and substations, effectively "smartening" these. The promise, to date fulfilled, is that this fusing of smart technology to the transmission system will help ensure that a complexity generated blackout will never again so easily propagate on our common wires. One result of this is that, according to a 2014 White House report, roughly 90 percent of power outages in the United States now start on distribution systems that have not been the object of the same care or recipient of cash.

There is another complicating factor. When it comes to new investment in long-distance, high-voltage lines, cost is not the only problem. We average Americans also pose a problem, for two reasons. First, we don't want to pay more money for our electricity, even though we enjoy some of the cheapest power in the world. Second, lines have gone out of vogue. No one wants them near their house. No one even wants them in their wilderness areas. No one wants to see them running along hilltops or through canyonlands. No one wants to cede their back forty for the passage of these great cords of modern life. "Not In My Back Yard" has come to the bulk power industry, and as much as we might desire reliable electricity delivered to our homes, we don't want to pay more for it, and we fight tooth and nail (as battle after battle has shown) to keep any new transmission from becoming a visible blight upon our land.

Last, money doesn't flow toward a critical part of the grid system— vars. Vars, or reactive power, help to stabilize voltage levels on the grid. Even alternating current, with its remarkable capacity to be transmitted so far without diminished force, needs occasionally to be buoyed up and smoothed out. The increase in long-distance wheeling after the Energy Policy Act has made maintaining power quality an even more critical element of basic grid functionality. This is what vars do: they help ensure a constant voltage in times of stress. As such they are essential to the well-being, reliability, and efficiency of the grid. They are, however, also a very hard sell, because strictly speaking they don't exist.

Vars are even more problematic non-things to understand (if not to make) than are watts. Nevertheless, before the act brought economies of profit to the grid, every utility ran some generating plants that made watts and fewer, far fewer, that made vars, because vars are what keep voltage and current in sync. With the separation of money made by generating electricity and that made by wheeling it everybody who used to make vars was, after 2000, trying to get into the watt market. Var production grew exceedingly slack. This lack of vars on the grid has been called "a key factor in the great Northeast blackout of August 2003" as well as in a series of blackouts on the Western Interconnection in the summer of 1996.

The return of the electric grid to the melodic strains of profit has meant that the companies that purchased recently divested power plants, as well as those investing in building new ones, are for the most part, uninterested in the var market. First because they can't sell them, and second because once the power they can sell is out on the lines its quality is no longer their concern. Problems like power surges (when voltage leads current) or brownouts (when current leads voltage) normally happen in the transmission of electricity, not its production. Producers don't need vars, and utilities, no longer in the business of producing power, reactive or otherwise, do need them to keep things stable on the lines. They don't, however, want to buy them because they can't be resold or otherwise rendered a viable product; there is no futures market in vars, for example. Consumers neither know, nor care, if the vars are present or if they've gone missing—until the power surges begin. The link between these surges (and blackouts and brownouts) and the absence of an invisible non-product from the electric grid is a simple reality about which most of us are entirely clueless.

The Eastern Interconnection careened out of control that hot August day not just because of what FirstEnergy did or failed to do. A bigger factor than those too-tall trees, that computer bug, is how the Energy Policy Act drastically changed the ways in which we now use the grid. The physics and the economics of the system today have no choice but to work at cross-purposes.

Meanwhile, we all also love what the act has made possible, if for different reasons. Lefty clean green energy types love it because at long last they have the right to buy power from less polluting, often more proximate sources, or even better, to make it themselves. The moneymen love it because they can finally make a profit off of the trade in electricity. The conservative right love it because it brings the free market, deregulation, and competition to what was a highly controlled noncompetitive industry riven through by the meddling tentacles of big government. Multinational corporations and multimillionaires love it because at long last they can invest in the energy sector with relatively little risk of loss. Building electrical generation, betting on electricity markets, making solar and wind mainstream technologies—it seems that whatever (social) way you look at it, the Energy Policy Act is a winning gig.

It may also spell the end of the utilities as we know them. It may even spell the end of the grid as we know it. But this is what reform looks like in America. Give the market an incentive and everyone with an idealistic glint in their eye will set themselves to working out how to make a better world out of opportunity, optimism, and grit.

What constitutes a "better world" is of course always up for grabs. But that's America too. If some people want to get rich while others want to put an end to global warming, well, let them duke it out in the marketplace. That's effectively what the Energy Policy Act has made possible. It's not a pretty fight. Nor is it a clean one. But where there was stagnation there is now innovation. And where there was a stable, if aging system—our grid—there is now a fantastically unstable, old one with new bits and more modern logics soldered into the joints and around the edges.

Be that as it may, at least the chips are in the air; the fight from here on out is all about how, where, and into whose pockets they will fall when they come back to ground.

CHAPTER 6

Two Birds, One Stone

In mid-July 2012, Thelma Taormina, a fifty-five-year-old woman from the greater Houston area, pulled a gun on the worker sent by her local utility to swap out her old analog electric meter for a new digital one. Like 2 million other Texans, Taormina was getting a smart meter that summer. Taormina, though, was uniquely displeased to see hers arrive. Her utility did nothing to help matters. The meter installer was rude, coming on her property despite a number of signs, clearly posted, proclaiming her dissenting position regarding smart meter installation. It was only after he pushed Taormina out of the way (she was physically blocking his access to the meter) that she got her gun and he finally saw the writing on the wall and left.

Taormina's complaint was that because these new electric meters wirelessly transmit real-time information about electricity usage they would allow the utility to not only track but also record and store patterns of her family's electricity use. "They'll be able to tell if you are running your computer, or air conditioner, whatever it is, they all put off different signatures," Taormina's husband pointed out in an interview with the local television station after the event. And though this sounds like paranoia, and the press has treated the Taorminas in precisely this far-fringe-of-the-American-right sort of way, the

Taorminas are actually correct. A German research team proved in 2011 that a careful study of the micropatterns of electricity use made newly observable by digital meters can reveal what appliances you have running and even what television program you are watching—as long as the program is being broadcast on scheduled TV, as signatures can be compared only with known possible outputs. Similarly, researchers at the University of Washington discovered that they could tell, based on electrical signatures, which houses were watching *The Lion King* and which were watching *Shrek 2*.

Even if no one ever goes through the trouble to monitor the Taorminas closely enough to notice that they cook more with their microwave than with their oven (this is conjecture), for them the smart meter is essentially a surveillance device, regardless of the utility's claims to the contrary.

The Taorminas are sure about this because the utility already has a perfectly serviceable machine for the purpose of measuring household electricity use—the meter being replaced. The analog induction watt-hour meter has been part of our electrical system since the nineteen-teens and still works just fine. Studies of meter accuracy—done by the utilities and also by the haters-of-utilities—have found both the old meters and the new meters to be sufficiently accurate, neither really outdoing the other in any substantial way. Realistically, the replacement of the analog meter by the smart meter cannot have been motivated by a desire to improve the accuracy of measuring electricity consumption.

Where the two differ is in the nuance of the information they provide. Digital meters offer a lot more data, which can, with the right software, be linked to time-of-day use, a very helpful data point for the utilities. And since they can be read remotely, they also allow the utility to cut costs by cutting employees. Far fewer people are needed to read the new meters than the old. Plus they get the utility off the phone.

Previous to the mass installation of smart meters utilities had to guess the parameters of an outage based upon triangulations of customer telephone calls. Before ten calls came in they couldn't even dispatch repair crews, since they didn't know the precise area affected. Now, with the new meters, they know if your power is out without you ever having to pick up the phone. Better still, they also know who among your neighbors has lost power, and who has not. This means that repair trucks of the right kind go to the right place right away. Since the introduction of digital smart meters, outage times have shrunk over all the United States.

But are these features enough to justify why so many utilities are spending so much and pushing so hard to make sure everyone gets the new meter? It was Texas's CenterPoint Energy's seventh visit to the Taorminas when the gun came out. From the Taorminas' point of view, and not theirs alone, the degree of importance the utility placed on getting them the new meter seemed suspicious.

What the Taorminas don't see, because it's been pretty well hidden from public view, is that something is in fact broken—it's just not the meters. The utilities themselves are suffering not only from a greatly attenuated revenue stream—which the new meters do help combat— but from a number of stresses that come from no longer being in control of the relationship between generation and consumption. Power plants from tiny to massive, many producing "variable" and thus unpredictable amounts of electric current, are popping up all over the place. Lines are more difficult to manage as well because of increased long-distance wheeling and, since 2003, increased automation. Arbitrage opportunities are shrinking because more smart people with proprietary algorithms have gotten into the energy trading game. And consumption is down after peaking in 2007, across the grid and across the nation. Despite the fact that we own ever more things needing to be plugged in, there is no indication that consumption levels will rise again anytime soon. The utilities, having lost, irrevocably, their iron grip on the electricity business, are being forced by circumstances to look elsewhere for stable revenue streams. Demand-side reform, it turns out, is one of the lone domains left to them.

In electricity speak, the "demand side" of the grid is the Taorminas. It's you and me. It's the people of Boulder, Colorado, and Bakersfield, California, and the 38.5 million other American households with installed smart meters as of 2014. Demand side means Walmart, and Google, and Alcoa Aluminum. It means the Old Executive Office Building in Washington, D.C., and all the lobbying firms on K Street. It means Appleton, Wisconsin. Up until now, other than having been encouraged to use more, more, ever more electricity, we "the demand side" have been almost entirely left out of the industry's considerations.

By controlling and monitoring our behavior as users of electricity, utilities can also control their own revenue streams. This is largely accomplished by trimming wasteful expenses on their side. Some of these might seem minor, though they add up, like shrinking the meter-reader force, or sending the right kind of repair crew to the right spot on the first try. Others are more momentous. Smart meters, for example, allow utilities to take steps to limit electricity usage at moments of peak demand—when we all ramp up our electricity use at the same time. Before smart meters, 10 percent of a utility's resources were always on the ready for these precise moments, though they only happen a couple of times a year and then often only for a couple of hours. The power plants reserved for periods of peak demand also tend to be the oldest and dirtiest plants in the fleet—plants that the utilities would like to see retired. As the people that breathe the air in America we also have good reason to want these plants decommissioned and shuttered for good. It's good that they are off most of the time; it's bad that they are on any of the time. But if these plants don't come online when needed, first brownouts and then blackouts will begin rolling across the landscape, as demand and supply fail to align.

To run these otherwise hibernating electricity factories the utilities need to make sure there is enough fuel sitting around ready to combust, to turn turbines that turn crankshafts that spin inside electromagnets that pulse vast quantities of electricity out into systems built to move less of it. They need enough trained staff on call and ready to go to fire it up and keep it running, and they need to ensure that it is in good working order despite sitting idle most of the time. This is massively expensive for them, and it is a simple and constant drain on their bottom line.

Finding a good way to control peak demand would offer the possibility of continued plant retirement, easing coal ever more thoroughly out of the number one spot for American electricity production and replacing it with smaller, more efficient, and relatively speaking cleaner natural gas combustion turbines (and occasionally even wind and solar power).

Likewise, the more electricity the lines are asked to carry the less efficiently they function. Peak demand doesn't just mean making more electricity with our least efficient plants. It also means losing a much higher percentage of that electricity in transit. This further lowers the amount of the potential power in a lump of coal that actually makes it into your reading lamp—down to right about 2 percent.

The utilities would thus prefer to control how much power we use or, barring that, charge us a lot more money for the electricity that is more expensive for them to make and wheel. In this way, they might recoup expenditures forced upon them by our proclivity to engage in the same power-hungry activities at the same times. It gets hot, we all turn up the air-conditioning; it gets cold, we all turn up the heat. In the UK, where electric kettles are common, they have a problem with something called TV pickup—a surge in demand during the ads of widely watched TV shows as television viewers head to the kitchen to boil up a spot of tea. This is a common enough phenomenon that England's utility often has three or four of these 400-MW surges to deal with every day. In most cases, the utilities have just rolled with these expensive behaviors. With smart meters, however, they finally also have the chance to pass this cost on to us, transforming peak demand from a moment of crisis for them, and for the grid, into a moment of peak revenue.

This interest in controlling customer use is brand-new. For the whole previous history of the grid the utilities did what was necessary to adapt to our whims, whether greedy or conservative. They did little, however, to openly intervene in how we the people used electricity. Consumption was a personal matter. Even when people wanted to begin conserving electricity by turning out lights, or insulating walls and buying better windows (and later, better refrigerators and clothes dryers), the utilities just grumbled to see their profits flatten as consumption dropped. This lack of intervention was also, effectively, a lack of communication. The utilities have a scant history of crossing the line that separates them and their operations from us and our dishwashers.

This is why utility intervention "feels" so weird, so bad, to the Taorminas and others protesting the mass deployment of smart meters. It's a sea change. The long-established laws of the electricity game have shifted under their feet—which are our feet as well—and, even without knowing it's happened, we can all feel it. In our collective gut we know the utility with its often graceless attempts at demand-side control has crossed some invisible line and is now, officially, behaving unusually. We are rightly suspicious of them, even if spuriously so.

It is true then, as the Taorminas fear, that the new meters are in fact information collectors that, once in place, can be used for communication about and control of customers. In many places, smart meters do already allow the utility to take remote control of your home air-conditioning and thus lower your "demand" by setting your thermostat a tad warmer (and so running the air conditioner a little bit less).

What the Taorminas (like so many other protesters) miss, however, is that at least for the moment these communication and control capabilities have very little to do with any sort of utility interest in the behavior of any individual customer. We matter to them as a manipulable aggregate, rarely as individual units, and almost never as particular people. This disinterest in customers as human beings with ideas, wants, beliefs, and proclivities that matter to them has been at the heart of almost all popular resistance to the mass rollout of digital smart meters, not just in Texas. Everywhere in the United States people were, to varying degrees, angry about this particular combination of forced intervention and personal disregard.

If the Taorminas are pretty sure that their utility's ulterior motive is to spy on citizens in violation of the Constitution and any number of laws of basic human decency, then the people of Bakersfield, California, also a highly conservative area, suspect the new meters are a tricky means of charging more for the same electricity. Their utility, PG&E, has to go through a series of complicated, highly bureaucratized procedures every time they want to raise the rates, so the smart meters that have magically and precipitously increased customers' bills without the normal hoop-jumping by the utility have raised some serious hackles. So ferocious was Bakersfield in its protests against the new meters that community-level resistance to meter installation, wherever in the country it has popped up, has become popularly known as the Bakersfield Effect, as in "The Bakersfield Effect Hits Santa Cruz."

A century of ignoring customers has translated into all kinds of flat-footed and tone-deaf interactions in the present. In Bakersfield, these were everywhere in evidence. From the customer surprised to discover that his electricity usage actually increased during a six-hour blackout to apartment dwellers who found themselves quite suddenly in the position of paying more for their electric bill than for their rent, smart meters looked like and felt like a cash grab. And though there has been a lot of quibbling as to why, nobody argues with the fact that with the new technology's arrival, monthly electric bills doubled, at times trebled. This happened in 2010, when Bakersfield had an unemployment rate of 16 percent and an underemployment rate double that. Nobody had $1,000 to pay a July electric bill that had cost $300 in June (not to mention $300 the previous July) with no appreciable change in consumer behavior. If both the old meters and the new meters were accurate, as the utility claimed, then the only viable explanation for the people of Bakersfield was highway robbery.

PG&E offered different interpretations: July 2010 had been much hotter, they claimed, than July 2009, resulting in more intensive air conditioner use; time-of-day rates meant that customers, not yet used to the idea that a kilowatt-hour of electricity might cost more at four in the afternoon than at six in the morning, were continuing to use power, and lots of it, at the most expensive times of day; and a poorly timed rate increase unrelated to the new meters had been rolled out as a means of covering the unexpectedly high costs of integrating more renewable energy into the grid.

As the Bakersfield protests grew in their vehemence, customers all over California started barricading the roads, refusing to allow trucks carrying smart meters to pass; baseball bats were seen (though no guns) as meter installers were run off of private property; and a class-action lawsuit was filed on the part of the citizens of Bakersfield. There were also some tinfoil hats, though those worries about electromagnetic radiation sickening customers who sleep too near the new meters were less common in Bakersfield than in Marin and Mendocino Counties.

Nationwide, Las Vegas County, Nevada, home to Sin City and surprisingly also to some of the most environmentally conscientious buildings in the country, became ground zero for radical fury as to the physical perils presented by the meters. In 2012 Las Vegas public utility commissioners were surprised to find an open meeting on the progress of smart-meter installation flooded by citizens claiming not only that they were being secretly monitored by their utility but that the smart meters were emitting rays that are the "rat poison used in energy." And though these protesters were not painting their homes with electromagnetic shielding paint (as has become common in Victoria, Australia, after their utility began the mass installation of digital smart meters), utility commissioners in Las Vegas soon found that their home addresses had been posted online, followed by the arrival of "strange packages" at several of the commissioners' residences. Meanwhile, in rural Maine, residents successfully pushed through an opt-out program because "safety standards for peak exposure limits to radio frequency have not been developed to take into account the particular sensitivity to eyes, testes and other ball-shaped organs."

The public outcry against smart meters has probably been loudest, and most resoundingly belittled, in the domain of health. The radiation emitted by the meters is very similar to that emitted by cell phones and wireless Internet routers, which while not proven to be safe are both types of electromagnetic pollution people opt to use all the time. It drives utility representatives a little bit more crazy every time they hold a hearing at which the very same folks there to complain about being poisoned by smart meter radiation are chatting happily away on their phones.

Consistency, however, is not the point. Control is. When, back in the day, limitless consumption was what customers desired, they felt in control of their electricity. There was always enough power available to them to do whatever struck their fancy. For the most part there still is. When conservation became a new value for America in the late 1960s, this sense of control was eroded as customers began to discover that it was very difficult to use substantially less electricity than they had been and even more difficult to see attempted changes in consumption reflected on their monthly bills. They also felt like there was little say they could have in how their electricity was being made, from what kinds of fuel, and with what environmental after-effects. Regulation since that time has had a profound effect upon the workings of the supply side of the grid, and even some ripples in the construction of more efficient appliances, but demand-side frustrations haven't changed very much. Customers still don't understand their bills, usage still doesn't seem to be linked to cost, and the promise of being able to choose cleaner, more efficiently produced electricity is held consistently at bay.

This is ever so slightly less true now. As I write in 2015, the tide has finally begun to turn. Today, one sees customer-driven initiatives to produce their own power, largely from home solar installations and community choice aggregation that lets municipalities decide the mix of fuels they want used to generate electricity for their town.

Some modest competition has emerged in larger markets between retail electricity service providers, which offer different types of plans to please different types of customers: 100 percent wind for a little more money, a flat rate for a little less money, and so on. Some of these changes have been almost accidental aftereffects of the widespread deployment of smart meters. The utility may have wanted to disconnect their capacity to do business from things like telephones and dumb, greedy air conditioners, but what they also enabled with these new meters is "net metering" by means of which homemade electricity, from rooftop solar panels, for example, can be fed back into the grid. The utility has to pay you for this power and the smart meter helps them keep track of how much they owe you at the end of each billing cycle. As of 2015, all the states except South Dakota, Mississippi, Alabama, and Tennessee had active net meter policies in place. In 2012, however, when smart meter ire was reaching its peak, these changes were not yet on many people's radar. All they saw was their utility once again doing things they didn't understand and about which they had no say.

Nowhere was this disconnect between what customers wanted and how utilities set out to manage them more off kilter than in Boulder, Colorado—America's SmartGridCity. Xcel, a Midwestern utility with an eight-state reach, decided to build a prototype of a future grid in a place where, they deemed, people would be receptive to high-tech means of reducing energy consumption. Boulder and St. Cloud, Minnesota, made the short list, and in the end Boulder was selected. Rather than just being given smart meters, the utility undertook, and vigorously promoted, a far more thorough upgrade. Their vision, as they tell it, was to build a real smart grid, a new grid of substance that could think, in many ways, for itself, with smart wires, smart appliances, smart thermostats linked in constant conversation with one another, with the utility, and eventually with home solar systems, battery storage, and a grid-enabled electric car.

The idea behind this first serious attempt to operationalize a real smart grid was an admirable one. Grid operators and a great many utility men understood that the revolution in telecommunications that followed from the shattering of Ma Bell in 1982 provided an interesting model for how electric companies might transition from government-regulated monopolies to customer-oriented service providers. No one could have imagined at the close of the 1970s (when most people still rented their phones from their local monopoly phone company, few people had long distance service, and those on a budget shared a phone line with some other random customer) that in thirty years' time phones would be tiny computers totally unlinked from the wall, and that absolutely everybody would have one of their own. The number of landlines in the United States has been plummeting since 2003, but this shift away from embedded infrastructure is almost irrelevant to the phone companies that survived the initial shattering. The survivors now have two new products to sell: Internet service and smart phones.

Though there was a lot of chaos in the transition and a number of phone companies did crash and burn along the way, there is something hopeful to the electricity reformers about the way that the phones themselves have survived as a product and how providing the interconnection between these phones has survived as a service. The grid, they hoped, could do the same, though perhaps on a more modest scale.

This idea of a transition like the one the telecommunications industry had just survived, if ideally with more winners and fewer losers, was not merely industry conservatism. The grid is the wires. These could be "smartened" (which is to say, rendered capable of transmitting information as well as electricity), while elements of the grid's communication structure could be rendered wireless—most notably meter readings; however, attempting to unplug electricity from the grid, as the phone has been unplugged from the wall, was too radical a leap. Even in 2012 the so-called smart grid was, just like our regular old grid, premised upon a robust endoskeleton of long-distance transportation and shorter-distance pole-top distribution networks: namely, wires. Whatever else happened, it really did seem like the vast majority of these wires were here for good.

In 2015, three short years later, we're not so sure. Once a technological revolution gets rolling it has the potential to outstrip even the most radical predictions regarding its end point. We might, in other words, lose a lot of wires in the coming decades. Not because we invent our way out of the need for long-distance transmission capacity—efficient wireless transmission of high voltages does not seem to be near at hand—but because distributed generation is bringing electricity production much closer to home. We aren't (yet) inventing our way around the wires, but we are changing our habits to make them less necessary. However, back in 2008, when Xcel started construction of its SmartGridCity in Boulder, this notion was so far beyond the pale as to be unimaginable.

The first thing Xcel set out to do in Boulder was very similar to the first thing Edison did in Manhattan: they started by digging up the streets to lay cable. In the end, this decision would be the one that effectively bankrupted the project; it turns out that Boulder has a lot more hard rock nestled just up underneath its pavement than anyone had predicted, or budgeted for. Etching conduit tracks into the stony underpinnings of the city proved both horribly expensive and time-consuming. In 2014, when the project officially ground to a halt, only a fraction of the total necessary fiber optic and broadband-over-powerline (BPL) cabling had been put to ground, at a cost of $21 million. As early as 2012, the company had admitted that BPL was an overly expensive technology that it would not use again.

With the benefit of hindsight, what is perhaps most interesting about the Xcel project was the degree to which a wireless future had not yet permeated the minds of the very men planning and constructing the city they hoped would be the cutting edge of this future. It seems inconceivable now that anything other than an Internet-connected wireless system for information transfer could even be considered. For Xcel, at the time, things were not so cut and dried. The grid, the sound logic went, was developing into a *wired* system for the multidirectional transmission of both information and electricity. It would rival existing telecommunication systems at the level of infrastructure while simultaneously transporting and distributing electricity from where it is made to where it is used. The smart grid was a real opportunity to upgrade the wires as well as a first chance to substantially alter customer usage patterns as a part of the long-term project of balancing and maintaining the profitability of the larger grid.

The first critical difference is that with a perfect, future smart grid, electricity might be made everywhere. Flows of power would be multidirectional as rooftop solar met backyard wind met big nuclear or hydro, and the output of all these types of generators would intermingle "intelligently" on the wires. Second, information would also travel freely between utility, customers, and the various machines that populate the grid, including both those that used electricity and those that produced it. The movement of both information and electricity would be monitored because meters could run both forward and backward, keeping perfect track of electricity outputs and inputs from even the smallest sources. Smart thermostats would form the hub of home-based networks of smart appliances that altered their behaviors without direct human intervention to take into account real-time, variable pricing of electricity. Since the lines necessary to carry radically distributed, uncoordinated, and small-scale generation to market also needed to be smart (which is to say to have some basic digital computing capacities), it made sense to build the two functionalities into the same set of wires. Information and electricity finally, in Boulder, becoming one.

In a perfect world, the various component parts of the Boulder smart grid would have functioned like distributed, systemic thought. The grid would make simple decisions from the dishwasher level all the way up to that of the power plant; it could communicate both to and between people and machines while balancing its load, all while avoiding the electrocution of its linemen, pretty much all by itself.

Perhaps Xcel, with its grand plans for Boulder's grid, overlooked the cheaper option of wireless information transfer (which almost all the nation's other utilities had grabbed on to with such vehemence) because their vision for both the city, and the future of the grid, was so much grander. They didn't just want to give Boulder smart meters, they wanted to make Boulder smart, at the most basic level of infrastructure, but also at the level of appliances. Xcel would smarten

Boulder's residents, too, so that they might interface seamlessly, happily, into all the radical technologies in deploy.

None of this materialized in Boulder. The fiber optic and BPL cabling was only ever partially installed, smart appliances never hit the market (they still haven't), only 43 percent of city residents got smart meters, and only about 100 households (rather than the planned 1,850) got smart thermostats. Even for those with the new meters and new thermostats, the promised "real-time readability" of electricity usage was neither real-time nor was it ever explicitly linked to cost.

Ski Milburn, an outspoken Boulder resident, captured the spirit of the project's many missteps in a 2013 interview: "I live in SmartGridCity," he said. "And I have a smart meter, and the problem . . . it's that the system is really, really dumb—so dumb as to be virtually useless. Case in point, the closest I can get to real-time energy consumption data is with a fifteen-minute delay. Over optical fiber! eBay can give me a millisecond response to something I'm bidding on halfway around the world and it takes Xcel fifteen minutes to give me something I could have gotten from my dumb meter by walking outside and looking at it. Give me a blinking break." Mike Warwick, also of Boulder, sums it all up (giving Ski the credit): "Ski Milburn hit the nail on the head. This was designed from the get-go as a 'Smart Grid/Stupid Customer' pilot."

In the end, it seems that the SmartGridCity failed to congeal because on many fronts Xcel executives failed to understand the spirit of the times. They laid wires when wirelessness, in all things, was the more popular path. They actually reduced customers' ability to understand their electricity consumption without contributing in any way to the one thing modern-day electricity customers want the most: the capacity to understand and control their own consumption and its cost, coupled with some measure of control over where and how their power is made. All the while the utility's constant meddling in houses, in streets, and with electricity rates irritated people who mostly wanted to be left alone.

In 2009, an executive from Xcel said, in utter exasperation, about the Boulder project: "We just don't know what our customers want!" Not five minutes before he had quoted a customer, who'd said: "All we really want when we get home is for the TV to work and the beer to be cold." It was almost as if Xcel couldn't hear what people were saying no matter how clearly these people articulated their positions on the matter.

The lack of conformity between the vision for the smart grid and the technology to make it possible turned out to be but one piece of the problem for the SmartGridCity. An equally destructive factor was Xcel's failure to communicate the purpose of the program to its customer base.

Much like the Taorminas of Texas, Xcel's Boulder customers didn't want the utility spending a lot of time meddling, or lurking about, on their property. They didn't want a more expensive and less stable electricity supply, they didn't want to interact with poorly made thermostats and smart home controller modules, they didn't want the utility remote-controlling their air-conditioning, they didn't want their streets dug up, they didn't want their dryer to turn on at two in the morning all by itself, and they really didn't want to be encouraged to do their vacuuming at midnight in order to save a nickel or two. In fact it was unclear from the customers' points of view what the entire undertaking was good for. As Boulder resident Stephen Fairfax put it in 2010: "How are meters that communicate with the utility supposed to benefit consumers? What does this give us that we don't have now? To many consumers the Smart Grid means that some bureaucrat will turn off their air conditioner when it is very hot outside." Everything that the utility cast in terms of choice and control was something that customers either didn't want or were spectacularly underwhelmed by.

It seemed that both the radical vision for the SmartGridCity and the reality of the halfway-done, sort-of-smartish grid the city actually got offered even less control to customers than they'd had under the old system. As a result, almost more than any other group of smart grid resisters in the nation, Boulder's residents rejected their utility's efforts on their behalf. They took their expensive utility-provided smart thermostats off the wall and put them in the kitchen junk drawer; they ignored their meters or refused to let the utility install them; they did nothing to limit or change their behaviors in relationship to restructured rates; and they grew furious as the project went over budget. When they were asked (along with the rest of Colorado's utility customers) to pay for these cost overruns, they lobbied the Colorado Public Utilities Commission to be freed from a burden they'd never signed up for.

And they won. Though Xcel's 1.7 million Colorado customers did have to absorb $27.9 million in cost overruns for the project, when the utility asked for $16.6 million more, the commission said that it would not consider a second bailout until the utility could prove they had provided some benefit to customers with the undertaking. To date, Xcel has not been able to provide sufficiently convincing information in this regard.

The question remains the same in Boulder as in Bakersfield and Houston and Maine (with residents' worries about the well-being of their ball-shaped organs): If smart meters or even a whole smart grid can't be proved to benefit customers even by the very utility undertaking the upgrade, whom, then, do they benefit? Why did Xcel go to the trouble and expense of building a citywide smart grid? Why did CenterPoint visit the Taorminas seven times in attempting to give them a smart meter? Why did PG&E risk a class-action lawsuit to ensure that all the people of Bakersfield also got their new meters? The answer, of course, is that smart meters don't benefit us, the customers. At least they don't directly. Smart meters, and to a lesser extent other grid-smartening investments, benefit them, the utility companies.

The level of expense and of risk for the utility companies is only suspicious until one realizes the stakes. The utilities vastly underestimated the degree to which reforming the industry would negatively impact them. By 2014, as Xcel was pulling out of its SmartGridCity commitments, as Colorado was withdrawing its fiduciary support of utility overspending, and as the people of Boulder were voting for a municipal takeover of their grid, there was a fatal phrase in the air: "the utility death spiral." It's melodramatic, but nevertheless suggestive.

America's for-profit utility companies are fighting for their continued survival; a number of them, including Texas's Energy Futures Holding Corporation, are actively dying. Some observers claim that what we are witnessing is not an evolutionary trimming of the business-model tree but an extinction of the utilities as such— permanent and total—such that in a generation kids won't know what the word "utility" means any more than they understand why we use the term "dial" for making a phone call. Others, including Warren Buffett, think that the notion of the utility death spiral is sheer fearmongering, and though some utilities will surely perish, as happens every time an industry undergoes a major transition, many of the utilities alive today will emerge from the singularity to bill us on the other side. Most likely the companies left standing will be the biggest of the big: Duke Energy, PG&E, ConEd, FirstEnergy, and probably even Xcel. Almost no one is optimistic enough to say that they will all come out of the transition unscathed.

Despite the moderate, modest, and often quite hopeful ways that grid reform began in the late 1970s, and despite the more substantial ways that the utilities had power and control wrested from them by the 1992 Energy Policy Act, most companies suspected they could weather the shifting seas by simply running a tighter ship. Through the California energy crises of 2000–2001, through the 2003 East Coast blackout, through the rise of variable generation and the forced privatization of generation in many markets, they believed that something like business as usual might persevere through the rough times. People in America would always need electricity, and who better than they to continue managing its provision?

What they didn't see coming was grid defection. For so long, "off the grid" had been a suspect term. Going off the grid was something that longhairs ingigliated in clouds of pot smoke and right-wing alligator farmers did (both groups form a notable presence in the new Colorado). It was not something that would ever occur to regular folks, city folks, and reputable mainstream companies. Not since alternating current did away with private plants in the 1890s, and not since the consolidation of power in the 1910s brought whole regions under the purview of a single service provider, had it been fiscally sound to be electricity independent. Then, quite suddenly, it was.

By the time Boulder's smart grid was being run into the ground by its inhabitants' sheer recalcitrance, a house anywhere in the country without grid-provided electricity had stopped being regarded with suspicion. It was now a "nanogrid"—a pretty cool-sounding sort of thing—while a larger installation, say, a prison or university campus, that could island itself from the big grid and take care of all its own locks and lights, central heating and cooling was neither radical nor suspect. In fact it was just the opposite; as a "microgrid," these institutions could make mutually beneficial deals with their utility to, for example, stop drawing power from the big grid during heat waves in exchange for substantial cash payments. Using available local generation—like solar panels, fuel cells, and diesel generators coupled with battery storage—these institutions need not pass privation along to their inmates and employees (in the first case) or students, professors, or researchers (in the second). All they have done is change where they get their power from, and when it's done right, nobody should even notice the change. As we entered the second decade of the twentieth century, the electricity business was afresh with opportunities. These opened whole new doors for companies quick on the uptake while leaving the less savvy to founder in the wake of change.

This transition happened so quickly that in 2008 the men behind the SmartGridCity could not grasp what was coming. In one case, however, Xcel not only anticipated this future of the nanos and micros and vehicle-to-grid storage, but they actually built it and gave it away.

This case was that of the home of Bud Peterson, then chancellor of the University of Colorado, and his wife, Val. Bud and Val, who'd been picked by Xcel to serve as the poster children for the SmartGridCity as a whole, had a much more thoroughgoing "smart grid" made available to them—one that in its details reveals what it actually takes to transform the electrical infrastructure we have into the electrical infrastructure we'd prefer. The dream in this case was not Bud and Val's, but Xcel's, which had imagined and compiled (almost) all the necessary ingredients for a real smart grid upgrade. In the process, they also very nearly got the Petersons off the grid.

Like many Boulder residents, the Petersons were given a smart meter and the broadband-over-powerline cabling that allowed this meter to communicate with the utility. They, and ninety-nine others, got the smart thermostat and smart wall plugs that could be remotely controlled or programmed via the Internet. For the Petersons, however, it didn't stop there. Xcel also installed solar panels on the Petersons' roof, gave them a hybrid electric vehicle (that's right, their utility gave them a car), on-site battery storage, and the more-or-less-real-time feedback to make it all work together. They didn't get the promised smart appliances, nor was the link made between realish-time electricity use provided by their meter and the actual price they would pay for that electricity. These elements were never incorporated into the Boulder grid, not even for the Petersons. These remain, however, critical elements of contemporary imagining of a future perfect grid.

In exchange for what they did get, the Petersons spoke publicly and glowingly in favor of the project. Said Bud, "Today, when you go online to buy an airline ticket, you can select on schedule or price. With this type of system, you will be able to select whether you would like to use renewable energy, sustainable energy, or coal-fired."

Said Val, "I pretty much get on my computer, tell my house and my car what to do and then I walk away. My solar panels are talking to my house, are talking to my car, are talking to my house. It's a beautiful system."

Said Xcel, "Since the Petersons started the program, they have been able to produce 590.7 fewer pounds of carbon, saving enough to microwave 154 pizzas. Multiply that by fifty thousand customers—the number currently expected to install the system—and it can make quite an impact."

No matter how much carbon the Petersons might have saved with their Prius and panels and battery packs, it will never be enough to microwave a pizza—not even one pizza—because running a microwave oven on carbon is about as feasible as flying an airplane with coal or running a hot water heater on bananas. Carbon is not a power source. I point out such factual troubles because, in general, the places where facts get wobbly help us, as readers of interviews with "satisfied customers," understand not so much what they have and what it does but what they'd like to have and what they would like it to do. As such, these interviews can be very revealing.

Bud, for example, thinks it's valuable to be able to choose what fuels are being used to power his world, and he expects that this is something a smart grid should be able to help him do. This, critically, was one of the main complaints that all Boulder had about Xcel, and not just in regard to the SmartGridCity project. They really wanted more renewable energy integrated into their electricity supply, while Xcel seemed wedded to the continued support of a nearby coal plant. When Boulder divorced Xcel in 2014 as the result of a citywide referendum, leading to the municipalizing of all local electrical infrastructure, the main reason given by voters for their vote against the utility was a lack of choice regarding generation. Whatever they thought about the SmartGridCity, what made them really angry was the utility's unwillingness to integrate more wind power.

Val, however, has different concerns. She wants her house to manage itself. She wants it to make electricity, store it, and use it without her having to do much more than punch into her smart phone, DISHES WASHED BY 5 P.M. and MAKE SURE THE CAR IS CHARGED BY 7.

The coming "Internet of Things," of which smart phones, smart appliances, smart meters, and electric cars are all integral parts (it is coming, by the way, it just hasn't quite arrived yet), is in many ways the continuation of an emancipation project that began in the 1930s to free women from the drudgery of household work by electrifying common appliances. The laundry line became the electric clothes dryer, the washboard became the electric washing machine, the icebox became the refrigerator, the kettle on the stove for the weekly bath became the electric hot water heater, the mangle became the electric iron, and so on and so forth.

The pleasure Val takes in her home's ability to "talk" to the various household machines without much intervention on her part points toward the next step in this project of emancipation. The machines will manage themselves. No longer does a woman need to be physically present to press a button or turn a dial. The solar panels will make electricity when then can, they will store that electricity for use as needed, and the storage device, usually banks of dedicated batteries or, even better, the battery in the car, will communicate with the smart meter to determine when the autonomous vacuum should slide out of its docking port, when the washer should click on, or when ice should be made in the rooftop freezer (later melted to cool the home, thus doing away with air-conditioning altogether).

Money will be saved, since grid-provided power will be accessed only when it is cheapest, and time will be saved, since Val Peterson will no longer need to be home to push the buttons.

This shifting of when everyday devices do their work from the time somebody is home to turn them on to the time that electricity is cheapest is not just appealing to Val; it is at the heart of why the smart grid is valuable to utilities. The reason companies would pay to outfit a house along the line of the Petersons'—and why they would dream of also including all the smart appliances the Petersons did not get— is because the smart grid makes it possible to shift consumption around the clock. It doesn't reduce consumption; in fact, quite the opposite: the utilities would be pleased if everyone used more electricity than they currently do. Rather, what the smart grid does is change the time at which consumption takes place. Instead of most of the nation vacuuming, washing and drying laundry, cooking dinner, watching TV, charging their car, and turning up their air-conditioning (or heat, depending upon the season) at the same time of day, some people, like Val, will let the house decide. And its decisions will be based upon time-of-day tariffs that cause electricity to be cheapest at night—when there is the least demand—and most expensive between five and ten P.M., when demand is highest. Val's house is very likely to use grid-provided electricity mostly at night; it's not happenstance that this is exactly when her utility needs it to be used.

The "smart" grid uses computers to alleviate the abiding problem of peak load. It also has the benefit of producing new things that customers might find appealing to buy: autonomous vacuum cleaners, programmable clothes washers, and learning thermostats. Following the model provided by the telecommunications industry the utilities would like to remake electricity into a new, more easily graspable commodity while remaking themselves into providers of services and gadgets. If they can manage this double task they will stay alive. If they fail a great many of them will very likely die, as the telecommunications companies Covad, Focal Communications, McLeod, Northpoint, and Winstar died during the deregulation of that industry, as WorldCom died in 2002—at that point the largest bankruptcy in U.S. history.

The success of the smart grid, therefore, has very real stakes. For utilities, it could stay their demise. For consumers, it offers a way of keeping something like a big grid and the equal access to affordable electric power that this enables. Smart meters, as the devices that at least in theory make real-time electricity use calculable, make the considered use of electricity possible for the first time ever. It doesn't quite work yet, pieces are missing, and the software is troglodytic, but with luck and creative diligence it will all come together in the very near future.

In giving the Petersons their SmartGridHouseCarComboPack, Xcel actually did something remarkable: they built one of the first mainstream instances of exactly the kind of nanogrid that we are still, a decade later, dreaming of operationalizing en masse. The Petersons, thanks to their utility company, were rendered very nearly self-sufficient. Had Xcel given their other fifty thousand Boulder customers an electric car, some solar panels, and banks of batteries, they would have changed the electrical profile of the city in a radical and unprecedented way. The vision of Boulder presented in 2008, had it been truly built, would have borne a strong and abiding resemblance to a future Phoenix (ca. 2029). Xcel would also have bankrupted itself.

This is in part because electric cars, solar panels, and giant banks of batteries are expensive, and in part because with all of these elements installed in their homes, very few people would have needed to buy their power from the utility. The big grid would have become a backup system. And the utility, which stakes its livelihood on the continued presence and relevance of this grid, would have been rendered obsolete.

This, then, is exactly the problem. The utilities don't know how to upgrade existing technology without putting themselves out of business. Nor do they know how to continue with the existing infrastructure without going out of business.

As a compromise, most utilities in the country are opting to install the smart meters but not to provide consumers with the rest of what the Petersons got. At least not on their dime. A smart meter allows for smart appliances, such as an air conditioner or thermostat, that can be set in conjunction with a utility's promotional-rate structure to encourage time-of-day use that smooths out peak demand. It also turns out, to almost nobody's surprise, that smart meters allow the utility to remotely control electricity use.

Despite all the chatter to the contrary, when it comes to being too hot or too cold, people tend not to act in economically rational ways. They follow a physiological rationality instead. This dictates turning the air-conditioning way up on the summer's hottest days and the heat way up to fight off the winter's fiercest cold. With a smart meter, many utilities can regulate these electricity-intensive proclivities before you, and all your extra-hot or extra-cold neighbors crash the grid.

The smart meter is the only part of the SmartGridHouse-CarNanoGridComboPack that is actually necessary to the utilities, because, to borrow the words of the technology journalist Glenn Fleishman, "shedding 5 to 10 percent of their load at peak times on demand could reduce or eliminate turning to the expensive spot power market or powering up dirty old power plants. Shaving that usage can have enormously disproportionate cost and environmental savings."

Peak load is the utilities' nightmare; it happens once or twice every year, and it's always a scramble to make sure things don't go disastrously awry every single time. This is why it would be useful to the utilities if people have smart meters that can both monitor usage and be used to control it in times of systemic stress. It is also why utilities like Consolidated Edison in New York robocall pretty much every one in the five boroughs on particularly hot summer days asking residents to pretty please turn down their air conditioners else risk setting off a citywide blackout. Many people rightly suspected from the start that "smart meter" actually meant "device for remotely controlling residential air-conditioning" even back when the official line was that it wouldn't ever be used to such Big Brotheresque ends.

The utilities' panic is real; it's not aesthetic, not even greedy, and not particularly malicious. As improbable as it might seem, it's real structural, organizational panic.

Most of the time customer usage is fairly predictable. Refrigerators, for example, are always on, and use about 14 percent of domestic power. Freezers use another 4 percent. Based on this data, the utility can bank on needing to make and transport a certain baseline of power all the time. This is one of the advantages of nuclear power plants. They may be difficult to ramp up or down quickly, but since they provide 20 percent of American power with exceptional reliability, they are basically big machines for transforming plutonium into refrigeration.

Other things, such as lights—11 percent of domestic power, but 26 percent of commercial electrical consumption—go on at a predictable point in time. The utility charts sunrise and sunset, opening hours and closing hours. They have a good idea of who is going to need the lights on when and where. Even though this is a semi-shifting point of data, utilities adapt their electrical production to meet demand, ramping power plants up and down at a slightly different time every day as summers slip into winters and winters back to summers again.

Still other things are culturally predictable. Between five and six P.M., Americans tend to come home from work. When this happens we use all kinds of electrical devices we weren't using before: big TVs, microwaves, washing machines, and garage door openers. On average we open our fridge nine times in the hour before dinner. All of these things add up to a fairly substantial jump in demand from just after the close of the workday until about ten P.M., at which point demand begins a slow downward slide that ends at four A.M.—the hour of minimum load.

To meet this steady bump in demand, power plants ramp up everywhere in America just before five P.M. Unfortunately, it is also when the wind tends to slow down and, at certain times of year, when the sun starts its setting, making renewables without backup storage the *least* useful means of producing power at the most necessary moment of the day. Because we still lack a good system for storing renewable power, America's evenings are powered by coal, natural gas, and the ever-present baseline of nuclear.

All of this is, again, pretty predictable from the utilities' point of view. They provide electricity to the same community year in and year out and they know people's habits—at least in a mass and calculable way—and they can plan for variations in load as long as behaviors are both predictable and moderate. "Balancing" load is for the most part a dull and nondramatic business. People use a little more electricity; the utilities produce a little more electricity. Demand is usually within limits the utility can meet. In fact, 98 percent of the time this is the case. The problem is that 10 percent of the utilities' resources are devoted to the other 2 percent of the time.

The few days a year that are causing all the problems are the reason utilities like Xcel, CenterPoint, PG&E, and ConEd are paying so much money to wire up their service districts with smart meters. These are the days when demand is neither modest nor predictable.

As American homes get bigger, as population shifts toward the southern states, and as air-conditioning becomes ubiquitous in all new construction, the summer peaks have grown steadily graver than those in winter. Some blame can also be apportioned to global warming: thirteen of the fifteen hottest years on record (since the start of record-keeping in 1880) have been since 2000, with 2015 being the hottest year ever recorded. Even though summers are getting hotter, driving up air-conditioner usage nationwide, there have also been some spectacular cold snaps as well, the polar vortex in 2014 being among the most memorable in recent times. All of these extreme weather events, including entire years that are hotter than they should be, cause more crisis points for the grid. If only the rich kept cool, then peak demand would be just another day at the electricity factory. But given that even Inuit in tiny villages in the Canadian Arctic now own air conditioners, we can give up hoping against hope that the sweltering masses won't have access to the one machine that will ease their suffering and soothe their woes.

The air conditioner is here to stay. And, as Fleishman, quoted above, succinctly points out: "Anywhere there's air-conditioning, smart grids will likely prosper." This is not just because these devices use a lot of electricity (they do), but because everybody uses them at the same time and because when it's very hot outside the utilities are already having a difficult time for a variety of reasons: long-distance wheeling goes up, spot markets get expensive, and lines sag and grow less efficient.

Add to this that the utilities are in fact running a much tighter ship than ever they did. "In the need to stay competitive," says industry expert, critic, and innovator Massoud Amin, "many energy companies and the regional grid operators who work with them have been 'flying' the grid with less and less margin of error. This means keeping costs down, not investing sufficiently in new equipment, and not building new transmission highways to free up bottlenecks."

Amin, a passionate man when it comes to questions of energy infrastructure (he saw the good that it can do as a child in Iran), puts numbers to all of this: "About twenty-five years ago, the generation capacity margin, the ability to meet peak demand, was between 25 and 30 percent—it has now declined to less than half and is currently [2008] at 10 to 15 percent."

What this means (and here I'm just translating) is that twenty-five years ago, in the early 1980s, we as a nation could increase our electrical consumption by about 30 percent at the drop of a hat, and the utilities had all these power plants sitting around primed for precisely this kind of sudden increase in load. They fired them up, blasted some fuel in, set the crankshaft a-spinning, and the electromagnet too, *fwamp fwamp fwamp* (but much faster than that at 3,600 rotations a minute) churning out a flood of electrons. As these electrons bumped one into the next the force of their serial desire flew down the wires and out to the switchyard, where its voltage was ramped way up and then shot down America's electrical superhighway at 60 cycles per second. A second or so later it is in your home helping your air conditioner cheerily blow its cool all over you and yours.

With such systemic "shock absorbers," as Amin calls them, in place, no day was ever too hot, no peak load too pointy, for the grid's infrastructure to absorb. Granted, there were fewer people in the United States twenty-five years ago, fewer air conditioners, and even fewer disastrously hot days. But even given these changes, had utility investment in infrastructure kept pace with population growth and GDP growth, peak-load days would simply not be the sort of panic-inducing and blackout-causing affairs that they are today.

In a way the utilities are right in paring back. Having 30 percent of one's power plants just sitting there 359 or so days a year, and 30 percent of the carrying capacity of one's lines equally left unused "just in case," is wasteful in all kinds of ways. The issue was not the tightness of the ship, or even the desire to shave and shave at the margin of the system. The problem is peak load itself. The phenomenon is what needs to be done away with, and without a viable means to store electricity the only way to control it is to control the part of the system that creates it—us.

We make peak demand with our air conditioners (and less often with our heaters). This is why the utilities robocall, why they are installing all those smart meters, why they want control over home air-conditioning, and why they prefer we all vacuum at midnight. Each and every weird demand the utilities have made upon us since the 2003 blackout comes back to this single problem: they need a way to control peak demand. At issue is not how much electricity we collectively use but when we use it. Remember Samuel Insull's point: it is costing the utility the same amount of money to keep all their plants basically idle between eleven P.M. and six A.M. as it would be costing to run them full bore. They prefer that we pay them something for the electricity they are capable of producing all night long rather than using nothing during this time and then, worse, asking them to spend even more to bring backup power online to make extra electricity at five P.M. when we want more of it.

Until the utilities can figure out a way to shift our time-of-day usage, their "backup" plan, which erupts into action when peak load appears unpredictably and dramatically on the horizon, is two-pronged. First, as we already know, they fire up a massive coal-burning plant or a natural-gas combustion turbine that isn't used 98 percent of the time. It just sits there except for those six or seven days a year when it is not just used but is critical and necessary to maintaining the viability of the grid as a whole. The second approach, which is done in concert with the first, is to ask high-power-consuming industries to scale their consumption way down, to shut down completely, or to do something even more shocking—to transfer their load to diesel backup generators. On very hot days, or less often on very cold days, the utilities literally pay lumber mills and smelters, prisons and public schools, to stop drawing power from the grid. Most of these big consumers don't have microgrids. They are 100 percent grid dependent, and using less power means making life pretty unpleasant for their workers and other inhabitants.

In New York City, in 2010, sixty-five different facilities had enrolled in one such program. Reductions at Rikers Island alone that summer accounted for 5.2 megawatts pulled off the grid on the hottest days of the year, providing the Department of Corrections with an incentive payment of $100,000 for the savings it incurred. It's worth adding that it is easier to take something away from prisoners and other institutionalized persons (like high school students) than from suburbanites. Turning the air-conditioning off in a prison or in a high school (New York's other high performer in 2010 was LaGuardia High School) is, as they say, "picking the low-hanging fruit" off the peak-load tree.

More expensive and more socially and technologically complex are incentivized "opt in" programs like the one Pepco runs in Washington, D.C. These give modest cash payments to individual households, usually in the neighborhood of forty dollars a summer, in exchange for "letting the utility automatically control their air conditioners on the few summer days when system demand is highest."

Xcel has started something similar in Colorado. Their program is called the Saver's Switch, and it involves a small device that resembles a beeper of old, installed by the utility on the outside of your house, just next to the central air-conditioning unit. This device allows Xcel to wirelessly switch the air conditioner off and on as needed throughout the summer. The program is touted as being "100 percent free to you." Plus, like Pepco, Xcel pays you forty dollars for participating.

For the moment, these are programs customers can choose, or choose not, to be a part of, and their relative success in places such as D.C. and Colorado have led to more and more utilities trying them out around the country. The Open Source Initiative, an organization devoted to protecting open-source software, estimated that in the summer of 2012, forty-nine utilities in twenty-five states had programs such as Energy Wise Rewards (the Pepco undertaking) and the Saver's Switch, with over 5 million customers already having signed up to participate. In almost every case, the smart meter is what makes this voluntary ceding of control over household energy use to the utility possible. It is their primary weapon in softening peaks. And it's begun to work.

In truth, though the utilities really would like it if you decided to use less electricity at exactly those moments that it's most expensive for them to produce it, but they don't necessarily trust you to do so. Your interests (to stay cool) and theirs (to avoid peak load) are structurally most misaligned at exactly these same instants.

The smart meter thus functions as a single stone meant to kill two birds. First, it allows them to monitor and control *for themselves* your electricity use, a functionality that has already been operationalized in many markets. Second, ideally if not yet actually, it would provide you with sufficiently accurate information to enable you to monitor and control your electricity use for yourself. The best-case scenario from their point of view is not, however, someone remotely switching off a forgotten light with their smart phone while running to work, but rather the kind of programmable intelligent home area network (a so-called HAN) like the one Val Peterson imagined for herself. The house, or its computerized mind, should be left to make reasonable decisions based upon your desires, while also aligning with the electric company's most pressing needs.

That this has not yet happened given all of the things we can now do with computers is suspicious, though I suspect that this is indicative of a profound mistrust of customers on the part of the utilities. Here I am with Ski Milburn: I am not convinced that it's a computing or even systems complexity issue. Rather, it is entirely possible that when given real and significant control over our electricity we, the demand side of the grid, will not act in ways beneficial to the utilities' interests. In other words, once we understand how much electricity we are actually using and how much it is actually costing us, what's to stop us from disregarding our utility and its problems completely?

Most people either actively dislike their utility—often because of exactly their flatfootedness in the customer relations department on display in Boulder, Houston, and Bakersfield—or we feel neither one way or another about them. The utilities generate a bill, we pay it; there is little room for affection or care. Why, after all, should we have to worry about the problems they're having managing their admittedly unwieldy business? In many ways they are as right to distrust us as we are to distrust them.

If, then, in 2010 customers had reached the apex of their irritation with smart meters, by 2014 industry insiders had begun talking about the "utility death spiral" as if it was inevitable. As if, once the meters had been installed, there would be no way for established utilities to stop customers from taking precisely the kind of control they had been promised and then using it to kill the utilities by a thousand cuts. In many places, each such cut looks rather a lot like a solar panel plus a battery pack as people begin to chart their own course out at the edges of the grid.

NOTABLE PASSINGS

In 2014 the people of Boulder voted to municipalize their electrical infrastructure. They simply bought the wires and all the rest from Xcel and bade their digging, meddling, poorly communicating, investor-owned utility good-bye. Their main complaint: the utility had not given them enough choice; in poorly realizing their

SmartGridCity, Xcel had failed to offer customers anything like real control. Mostly the people of Boulder wanted the utility to ensure that more renewables and less coal was being used to generate power for the town. A second sticking point, however, was that if they were going to suffer time-of-day electricity rates, they wanted (as Ski Milburn pointed out above) the real-time information necessary to make informed decisions about both electricity use and its cost.

In the end the people of Boulder, Colorado were essentially guinea pigs. They were used by their utility as part of a PR campaign that was also a research project into grid reform. The SmartGridCity was no more about giving people substantive control over their electricity use than it was about giving them a smart dishwasher or a free electric car. Instead, what Boulder residents got were higher bills, a less reliable electricity system, and effectively no communication save a rain of glossy salvos from the utility.

Xcel survived, even as their SmartGridCity died, as an idea and in fact. Other for-profit utilities, large and small, have not been so lucky. If, from the end of the Depression Era to 1988, no major electric company in the United States filed for bankruptcy protection, then in 1988, as competition crept into the sector, the impossible began to happen: power companies started to fail. First one in New Hampshire, then one in El Paso, a small one in Colorado, and another in Maine, here and there they'd fold and quietly go under. The first really big bankruptcy filing was PG&E in 2001 (since restructured and revived).

Enron's collapse, also in 2001, pointed to another future for failure. You don't have to be an electric company to go bankrupt in the electricity business. By 2001 electricity was often controlled by larger energy companies—often investment consortiums—that might own a utility, some power plants, and maybe even some lines. Enron, with $65.5 billion in assets, was the fifth-largest bankruptcy in U.S. history; it owned three utilities, only one of which was in the United States

(Portland General Electric), and thirty-eight power plants worldwide, including ten wind farms in the United States and eight hydroelectric dams, all in Oregon.

The collapse of the Energy Futures Holding Corporation in 2014, a less well-known industry bomb than Enron, continues this trend. The eighth-largest bankruptcy in U.S. history, with $40.9 billion in assets, Energy Futures was only two-thirds the size of Enron, but all of its holdings were in a single state—Texas. Energy Futures Holding was the parent company of North Texas's retail electric service provider TXU Energy, which has more customers than any other electricity retailer in that state. Energy Futures also owned the state's largest power company, Luminant Generation Co., with twenty power plants across Texas (including a nuclear plant) and a generating capacity of 18,300 MW. And Energy Futures controlled Texas's largest transmission and distribution company, Oncor, with 7.5 million customers in more than four hundred Texas cities and towns, including Dallas–Fort Worth.

Energy Futures failed not because it was crooked; its bankruptcy was run-of-the-mill, utterly unlike Enron's. It made a few too many bad business decisions over the years. What is interesting about its collapse, however, is how evident it became as the business was stripped down for public viewing that even things that look like utilities aren't really utilities anymore. What's really there after deregulation, restructuring, and the incorporation of the free market is an investment company running a bunch of endeavors that when taken together were, once upon a time, known by the name "utility." Even if the "utility death spiral" doesn't get them, the market, it seems, already has.

CHAPTER 7

A Tale of Two Storms

For most people, December 1, 2007, was just a regular early-winter day—a make a cup of coffee, commute to work and home again, cook up some dinner, watch a little TV, give the kids a snuggle and pack them off to bed kind of day. Not so for Sylvie Straw and about a hundred thousand of her nearest neighbors scattered along eight hundred miles of Pacific Northwest coastline. For on that morning, one of the worst storms ever to hit the Northwest Coast rolled off the ocean and slammed into a dozen rural counties hours from anyplace important. They creaked and shattered under a wind that blew at 100 mph, gusting at times to 140 mph, until every community from Crescent City in California to Forks, Washington, was either cracked or flattened or flooded.

Imagine a wind so strong that windows curve inward with the force of each gust. Imagine those windows shattering after the fifteenth or twentieth such gust, an icy rain howling down and in, as the night of the first day gives way to the sickly yellow dawn of the second. That is how it was. On the second day, the storm continued unabated. Dormers were blown clean off houses. Trees hundreds of years old, some with the girth of giants, were ripped up by the roots and flung down on roadways, on people's roofs, on power lines, and on one another. Forests were laid flat. Every road in and out was rendered impassable. Every cell phone tower was down, every phone line dead. The isolation was total and it was not fleeting. The power was out for a week. The roads were closed and phones were dead almost as long.

Sylvie Straw, a then fifty-seven-year-old newly retired school-teacher, spent most of that storm—and the week of dark, cold December days that followed—huddled on an old couch in the back room of the nursing home where her husband worked. Sylvie is no stranger to isolation. Originally from the South Side of Chicago, she chose to live on this remote spit of land at the end of the earth. But this storm, this storm was another whole order of lonely: "We were all so isolated," she remembers. "It was so surreal, like going back in time; that was the feeling here in 2007. Like, 'This is the way it was before.'" She puts her coffee cup down with a bang. "And I really don't like it!"

When I spoke with the Straws it was April 2011, more than four years since the Great Coastal Gale, as the locals call it. We are sitting in the Straws' house, a minuscule, rain-cloud-gray Victorian, just a block from the mouth of the mighty Columbia River. Sylvie has invited a group of her friends and neighbors over to discuss the storm and its aftermath. There is a retired ship captain, a therapist and his wife—smart as whips and twice as quick—a social worker, the manager of the local chorale, an engineer, and a retired graphic artist, who made his living doing technical drawings for PG&E. He was the guy whose job it was to draw the poles with lines, transformer canisters, and glass insulators all to spec.

Everyone is balancing a plate of carrot cake on their lap and sipping at a strong, black cup of coffee as they chat. You wouldn't know by looking at them, but seated around me on an overstuffed flowery sofa and in a couple of matching armchairs are a bunch of radicals.

These nice, middle-aged, middle-class folks in seasonal knitwear are not here to talk about how they rebuilt and recovered from the storm of the century—though they did rebuild and recover. They are here to share how they gave up relying on utility-provided electricity.

Like many others the nation over, Sylvie Straw and some of her friends have taken the production of heat, light, and hot water for their homes back into their own hands. No one in this group is making all their electricity all the time by themselves. Rather, they have each cobbled together some sort of hybrid system that works when their grid doesn't. And, in certain cases, that also works alongside their electric grid—in concert with it. They have natural gas heaters and water heaters, propane-powered burners and lights, hand-cranked shortwave radios, rotary phones, and even a few diesel generators between them. They have made these choices because of the frequency of weather-related outages in their small point of land, the very last bit of solid ground before the wild expanse of the Pacific.

Not all of their ilk, the "Locavolts," to borrow a term coined by Tyrone Cashman (whom we met back in chapter 4), have made choices for electricity independence based upon their local grid's lack of resilience in a storm. All across the Southwest, the Northeast, and Hawaii, people are mounting solar panels on pretty much every south-facing thing that doesn't move (and even some that do)—houses, garages, office buildings, parking garages, empty fields, even RVs—while computer-intensive industries and electricity-reliant enterprises like all of Silicon Valley, many university campuses, military bases, prisons, and even the state of Connecticut are building microgrids that can be isolated from the big grid in moments of crisis.

These bigger systems rely on expert knowledge to plan, build, install, and manage, while in the case of the distributed solar revolution, third-party ownership (which basically means that the panels are leased from a private company that installs them on homes, garages, and other structures) has become the norm—75 percent of California's residential solar works in this way.

In the rural Pacific Northwest, however, there is still no blueprint for the systems being cobbled into place. Folks there can't order their get-off-the-grid kit from a catalog and assemble it on the living room floor. There is no IKEA of alternative energy systems, and thus each of the solutions they have individually come up with is different in its details and particular in its priorities. That said, there were some pretty strong opinions about how to best preserve heat and power and access to hot coffee when the grid was down. (The last of these was one of the prime issues under discussion. It turns out that the lack of hot coffee that miserable week was one of the storm's greatest offenses; curiously, in my discussions with Serbians who lived through the blackouts made by the 1999 NATO bombings of Belgrade, they also mostly complained about the lack of hot coffee and detailed the remarkably complex and often dangerous measures they took to acquire it.)

There are, according to Straw, good systems, honorable systems, for making one's own heat, light, and power. There are also bad ones. Neither batteries nor diesel generators were getting any points from her. After Superstorm Sandy, I heard very similar sentiments from stranded New Yorkers. Generators, batteries, and even electric cars are only as robust as their supply chain, and this, in a pinch, is rendered almost immediately as fragile as the system it is meant to buoy up.

When she was bivouacked in the nursing home during the gale, Sylvie Straw explains, she shared her couch in shifts for the duration of the storm with the man whose job it was to fill the generator. "He was exhausted," she remembers. "Even more so than the rest of us." She is explaining something that everyone (in her opinion) should already know. Diesel generators—the Pacific Northwest's go-to backup system for power outages—aren't any more reliable in a real emergency than utility-provided electricity. "The generator at the nursing home was huge because of people on oxygen and stuff. But how to keep it filled? Even with lights only on in the hallways and the temperature in the whole facility at sixty-three degrees, there just wasn't enough gas to run it."

Straw is angry all over again in retrospect: "There was nothing left at the docks. That's where you usually go to get gas, at the docks where the boats fill up, but those pumps, they were empty after the first day!" She remembers when Walter, the guy in charge of generator logistics, came in toward the end of the second day, exhausted, wet, and triumphant because he'd caught a glimpse of an oil truck. "He chased that truck down," she laughed, "imagine that. The wind is blowing at a hundred miles an hour, branches and trees are flying around through the air, and there's Walter speeding along trying to catch the oil truck! It took some fast talking on his part, that's for sure, but in the end the oil guy stopped by every day or so and filled up the generator."

Straw pauses for a moment (she is about to explain how the old people in the home are incorrigible cheats at bingo) and says: "That's what people don't realize. They think 'Well, I've got a generator, I'm safe, I'm warm, I've got light.' But no, what you've got is a problem, because when a real storm hits you don't have any gas for that generator. You're just as cold and miserable as the rest of us." Bob Johansson, her neighbor—eighty if he is a day—a quiet white man in faded blue overalls, chimes in: "Better you do it yourself."

"Bob saved my hide," Straw interjects; she is a fair measure more loquacious than her more modest neighbor. "I mean, I was freezing! We were so unprepared for that. And so I just went over to Bob's. He had heat; he had coffee. And he was showing me all this cool stuff he had, that he had compiled after the San Francisco earthquake . . . that shower was what just blew my mind." (Says Bob, "It was just a fertilizer sprayer that you buy at the hardware store with a shower head on it and you pump that up, get the pressure up manually, and then the water comes out.")

That's Bob. Practical. But that's because he'd lived through it before. This is something else I learned from these rural residents—something that everyone who suffered through similarly long, cold, miserable days in the wake of Sandy now also knows: once you've been through a really bad blackout and see how ill-equipped the system is to deal with disaster, it's tempting to take matters into your own hands. You have to go through enough storms or earthquakes or tsunamis or what have you—or you have to have gone through just one that was bad enough—before a "switch is flipped": electricity is no longer magic, it's a pain in the ass.

The San Francisco earthquake in 1989 did the trick for Bob, but he was quick on the uptake. Lots of people lived through 1989 and never thought to so much as buy a generator, not to mention design and build their own shower. Many more of those same Californians lived through the deregulation-generated blackouts and brownouts in 2000–2001 and still just hunkered down and waited for the government or the utility or some regulator some-where to put the system right again. Not the tech companies, though. Google, Apple, Yahoo, Cisco (and many others like Facebook, which didn't exist in 2000) can't lose power, not even for an instant. They are companies that brook no interruption. They can never be down, offline, or out of power. So, like Bob, they have built themselves some systems, different in each individual case, so that when the grid goes out, they stay on. They are warm and well lit; they have coffee, even probably some showers, and most important—computing power.

Some of them are using fuel cells or solar and wind with backup battery storage, but, like the Straws, fewer every year are relying upon diesel generators, and for the same reasons.

Sandy did something in New York similar to what the Great Coastal Gale did in the Pacific Northwest. People who had been considering a plug-in electric car stopped considering it. Most of the people who got out of that storm's blast zone after the fact did so because they had access to a car with a full-enough gas tank. They packed up that car and drove away, south usually, to friends and relatives anywhere beyond Sandy's disastrous reach.

It wasn't just individuals who felt their priorities change with the storm's winds and rains. In many ways, Sandy did for East Coast businesses and East Coast utilities what deregulation never managed: it made them realize that resiliency in a crisis needs to be built into the system, from the ground up. They realized that discussions about infrastructure are not only discussions about markets and money but also matters of survival. Sandy was New York's Great Gale. It was a storm that changed people's minds.

Meteorologically speaking Sandy was a far lesser storm. It was where and when it hit that account for its impact. Sandy, which smashed into the mid-Atlantic states on October 29, 2012, was a post-tropical cyclone wrapped in a Nor'easter "fully coupled to the jet stream." A triumvirate of force whose fierce wailing breath whipped up the waters at exactly high tide over the most densely populated 56,000 square miles in the nation. If the Great Gale, with its epic reach, affected 100,000 people, Sandy, feebler by far, hit 50 million. With maximum sustained winds at about 80 mph, the damage done by her wind was nothing compared with what her water did. Tides measuring almost fourteen feet high engulfed everything, sloshing into subway tunnels and up over coastal seawalls.

People were not just stranded, they were immobilized. For some, the lights were out for a month.

As in the rural Pacific Northwest, individual homes and businesses were battered by the superstorm. But for the most part, in the crowded East the things that were destroyed were those that make life in common bearable: the common electric grid, the subway and other transit systems, access to potable water, functioning sewers. It was a havoc-wrecker upon infrastructure. Just as the gale radicalized mild-mannered townsfolk too frustrated, cold, and isolated to ever want to go through it again on the same terms, Sandy put the fear of God most especially into those New Yorkers and New Jerseyites charged with keeping the lights on for everyone, the commuter trains running, the people warm, the companies churning out widgets, and ATMs spewing bills. Sandy made businesses and governments (state, local, municipal) reconsider what it takes to produce reliability under duress.

Similarly, if Northwesterners worked in the wake of the gale to connive and cobble together individual systems in response to their too-fragile infrastructure, Sandy has had a more profound impact on our thinking about what infrastructures mean for living life in common. In that storm's wake we began asking in earnest how this infrastructure might be built to work in the aftermath of disaster. If it can't be made to bend in a high wind at least, the new thinking goes, perhaps we can avoid its being so inexorably broken; perhaps we can rebuild it in such a way as to assure its ability to recover quickly.

On both the East Coast and the West, after both Sandy and the Great Gale, "resiliency," rather than "independence," has become the watchword for reform. Resiliency means the ability to take a blow and not be bowled over by it; it means designing ones and structures that can bend but not break; it means blackouts that bounce back into brightness rather than cascade across the continent; it means backup systems so seamlessly integrated into primary systems that one doesn't even notice the switch between them. Resiliency means accepting that sometimes things do break and then imagining and engineering ways not so much to make them unbreakable, as to consider how they might be less thoroughly broken in the first place and thus also easier to fix.

Resiliency is a different way of thinking about security than is usually taken after significant disasters, when the aim is to rebuild stronger, bigger, more solid systems (or structures) that can withstand the same stressor that felled their predecessors. That route, "The Hard Path" or "Infrastructural Hardening," is equivalent to managing illegal immigration by building a border wall, or protecting a frequently flooded area by strengthening and expanding a dike or a seawall. There is even a hint of the hard path in the initial years of the war on terror, when finding ways to gather total intelligence seemed the surest route to victory, regardless of the cost to human dignity and civil liberties. The hard path is a way of building security into systems that is premised upon both rigidity and mass—whether masses of concrete or masses of information.

This does not mean that the hard path is always a foolish one. Hardening can also mean converting wooden utility poles to cement or adding guy wires and other forms of structural support, which make these poles less likely to fall over in a high wind. Higher-voltage towers are usually made of aluminum, a weak metal that is not necessarily the best choice for areas expecting occasional direct hits by hurricane force winds. These are slowly being hardened as aluminum is replaced by galvanized steel or concrete. Substations are being moved to higher ground in easily flooded areas and distribution lines are being buried in areas susceptible to severe weather but not to flooding. Sometimes something as commonsensical as storing critical equipment on the upper floors of a building rather than at or below ground level makes all the difference when the lower eight feet of everything is soaked by a serious storm surge. All of these practical and affordable reactions to a vulnerable grid are part of the hard path. So, too, is the long-standing preference among U.S. utilities, regulators, and investors to increase the size and complexity of the grid's more critical elements and to rely more assiduously upon fossil fuels.

From the 1970s through the first decade of the 2000s the hard path, for the grid, was mostly about making power plants bigger and adding higher-voltage, longer-distance lines to more firmly interlock regional systems. It was also about relying on fuels that were both finite and hard to transport in times of stress: coal, oil, natural gas, and plutonium, as these were the most reliable ways to make massive quantities of a standardized electric current. As late as 2010, despite undeniable shifts in both the form and governance of electrical infrastructure, the hard path was still the primary way that improving the U.S. electric grid was figured by both the government and by the for-profit utilities.

It is also a way of imagining "security" that a solid subset of the American Left has been arguing against since the 1970s. Amory Lovins, who coined the term "the Soft Energy Path" in 1976 as a means of articulating an alternative way of thinking about national security, argued that our hardened infrastructure, far from making things stronger, in fact increases infrastructural brittleness and thus also the ease with which the systems that sustain us can be broken.

It is interesting to note that for the Lovinses (Amory wrote most often together with his wife, Hunter), security is not a domain isolated from the everyday. For them, energy security, oil supply chains, and a robust electrical infrastructure are all singular and inextricable elements of national security. Without our grid almost every element of modern life is lost. No computing power, no light, no telecommunications, no entertainment or news, no public transit, no air-conditioning, and in many places no potable or hot water. An otherwise undamaged refinery or pipeline will not operate without access to electric power. Gas stations dispense fuel by means of electric pumps, making supplies of liquid fuel for cars and airplanes also at risk when the grid breaks. Even money these days is more electric than it is material; there are no cash machines in a storm, no banking or investment systems, no ways to monitor, use, or control currency not made of paper or metal. No communications networks also means inoperative military and police forces.

In 1982 the Lovinses published a report commissioned by the federal precursor of FEMA in the wake of the previous decade's two devastating oil embargoes. Their conclusion: "The energy that runs America is brittle—easily shattered by accident or malice. That fragility frustrates the efforts of our Armed Forces to defend a nation that literally can be turned off by a handful of people. It poses, indeed, a grave and growing threat to national security, life, and liberty. This danger comes not from hostile ideology but from misapplied technology. It is not a threat imposed on us by enemies abroad. It is a threat we have heedlessly—and needlessly—imposed on ourselves."

"Our reliance on these delicately poised energy systems," they continued, "has unwittingly put at risk our whole way of life."

From the Lovinses' point of view, the power outages that occur regularly in the United States regardless of their cause—whether big storms or changing legislation or computer bugs or terrorist hackers—are a natural and utterly predictable side effect of having such a big, centralized electrical grid. "The size, complexity, pattern, and control structure of these electrical machines," they wrote, "rotating in exact synchrony across half a continent, and strung together by an easily severed network of aerial arteries whose failure is instantly disruptive . . . make them inherently vulnerable to large-scale failures."

The grid is awesomely complex. It is the largest machine in the world. To the Lovinses, however, the grandeur of this complexity is less impressive than it is foolhardy. If energy security is our goal, rather than simply electrification, such intricacy is a poor route to the system's reliability. We have seen this to be true in the 2003 East Coast blackout. The Lovinses made the same point twenty-five years previous, in a careful analysis of the 1977 New York City blackout. The conclusion could also have been reached in the wake of the 1965 Northeast blackout, caused by an incorrectly set relay just outside Niagara, and of the 2011 blackout in the American Southwest, which propagated from a lineman's error in Arizona, and of a 2014 blackout in Detroit—home to the nation's least reliable municipal grid—when a single cable failed.

There are storm-made tragedies, like those that followed from the Gale or from Sandy or from Hurricane Irene (which blacked out over 40 percent of the people living in Rhode Island, Connecticut, and Maryland in 2011), and then there are blackouts produced by complexity itself. The Lovinses, however, would argue that the severity and reach of any storm's damage is also linked to the intricacy and complexity of the electricity system in its current form.

For the Lovinses, reconceptualizing and then rebuilding our systems-in-common as smaller, more flexible, more self-contained, less polluting, and closer to home was the wisest way to proceed. Soft energy technologies, the adoption of which they considered to be the first necessary step toward ensuring energy security in the United States, have five defining characteristics. First, they rely on renewable energy resources, like wind and solar, but also biomass, geothermal, wave, and tidal power. Second, they are diverse and designed to function with maximum effectiveness within specific circumstances. Third, they are flexible and relatively simple to understand. Fourth, they should be matched to end-use needs in terms of scale, and fifth, they should also be matched to end use in terms of quality. All of this is nested within a larger cultural commitment to energy efficiency that is built in to our structures and our life-ways from the ground up. For the Lovinses, the soft energy path is not about privation but thoughtful, thorough integration of energy use into social life.

The first maxim is fairly self-explanatory. Renewable energy resources don't need risky supply chains—like the chance arrival of a fuel truck in the middle of a hurricane—in order to function. Because it is more difficult (though far from impossible) to disrupt their functioning they are by their nature more secure.

The second criterion needs a little more explaining. To say that an energy system should be diverse and designed for maximum effectiveness in particular circumstances is, essentially, the "don't put all your eggs in one basket" maxim of maintaining constant access to electricity. For the Lovinses, every power-supply system regardless of its size should integrate a number of technologically different sources of generation, with different weaknesses and different supply chain problems. A diesel generator is fine as long as it is only part of the mix. On a larger scale the same could be said of a natural gas turbine or solar array that is used to balance a wind farm.

The Lovinses would say, however, that one or two components are not enough. Never just diesel; not just natural gas or solar plus wind, but wind plus natural gas plus battery storage plus conservation and energy efficiency measures all taken as one. Diversification, as any investor knows, is the essence of a robust portfolio. The same, they argue, is true of power production, transmission, and distribution systems.

What is interesting about the Lovinses' formulation here is that they don't just imagine a cookie-cutter world in which the same four or five elements make up the electrical infrastructure for a Florida suburb prone to flooding, a research laboratory highly sensitive to small outages, a paper mill in the Rockies, a military outpost in Afghanistan, or an individual home in Maine, or Arkansas, or Utah. Rather, they urge that each set of circumstances be considered in light of the mix of available technologies that will best meet both the risk profile and the explicit needs of the end user while incurring the least possibility of failure. This leads straightforwardly to the third, fourth, and fifth elements of a resilient power system: it should be both flexible and relatively simple to understand, it should be matched to end-use needs in terms of scale, and it should be matched to end use in terms of quality.

From a study of biological systems rather than technological ones, the Lovinses argued that an organism's longevity consistently relies upon "local back-up, local autonomy, and a preference for small over large scale and for diversity over homogeneity." All of these things increase resilience in all cases. It's a strong argument, but one very much at odds with the prevailing modes of imagining and designing energy systems at the time of its publication. From the late 1970s until the first solid smack of wind delivered by Sandy thirty-five years later, resiliency, or the soft path, was just another radical thing that certain enthusiasts argued for with modest success (like building sustainable communities, eating locally sourced food, or practicing voluntary simplicity).

With Sandy, all of that changed. Resiliency as a well-developed philosophy of energy security stepped out from the shadows and into the spotlight. Even Republican politicians from the broad swath of storm-affected states are now advocating, and in certain cases mandating, that the grid be built and managed differently. They want small, "islandable" grids that rely upon a mix of resources to become the norm rather than the exception. This shift toward support for smaller, more local, more independent, and more resilient "micro" grids isn't necessarily a total embrace of all the Lovinses' suggestions. However, the two visions for a more secure world electric grid do have a lot in common, not least of which is the term "resiliency" itself. No longer a notion from the fringe, it is now the stuff of headlines, federal reports, and pontificating pundits.

The White House, for example, recently released a report called "Economic Benefits of Increasing Electric Grid Resiliency to Weather Outages." In it, they define a resilient grid as one in which "outages will affect fewer customers for shorter periods of time." This is to be achieved by a mix of "hardening, advance capabilities, [and] recovery/ reconstitution" in ways that integrate "high- and low-tech solutions." For the Executive Branch, resiliency is neither a hard path nor a soft path. More of a result than a process, grid resiliency will lessen our outages (and their impact) by making utility poles stronger and moving critical equipment to higher floors, by adding synchrophasors and smart meters, by manufacturing substations on American soil, and by adaptating the grid to more local needs.

Resiliency as it has been taken up by the states, the military, and many businesses has added another possibility to the federal laundry list for grid improvement—a possibility with a startlingly old-fashioned shape.

After Superstorm Sandy, the Northeast began to witness the return of the tiny grid. These new constructions bear a lot in common with Edison-era private plants, which generated customized electricity for a single owner on-site. Unlike Edison's private plants, these modern microgrids can connect and unconnect as needed to the big grid (which is now increasingly known as the "macrogrid"). And, unlike any system since the consolidation of power in the early twentieth century, these microgrids work perfectly well in "island" mode. When they disconnect from the big grid during a blackout, something most microgrids can do easily, if not always smoothly, they tend to stay on. In this way, they function as pockets of power, light, and heat in the midst of great swathes of distress.

During Sandy, for example, the microgrid at SUNY Stony Brook took over supplying power to the campus, which experienced a single nighttime hour of outage, in comparison with the ten cold, hungry days suffered by the surrounding area. Seven thousand students were housed in dorms in the main campus during the crisis, and many others were moved to the main campus from satellite campuses that had lost power. The campus police force and medical center were fully operational throughout the duration of the storm.

Similarly, the South Oaks Hospital on Long Island stayed on and fully functional during and after the storm. South Oaks, like SUNY Stony Brook and NYU on Manhattan, has a microgrid fueled by a cogeneration plant—a technology for making heat and power that dates back to the 1880s. Farther from home, "the Sendai one megawatt microgrid at Tohoku Fukushi University operated for two days in island mode while the surrounding region was without power" after the magnitude 9.0 earthquake and subsequent tsunami devastated Japan in 2011.

If in 2015 there were three hundred microgrids in the United States—currently the world leader in such systems—half again as many will be built in the post-Sandy years in the state of Connecticut alone. Usually smaller than 50 MW, microgrids are designed to be a lot like the smaller, more flexible infrastructure imagined by the Lovinses back in the 1970s. The main difference between their vision for resiliency and contemporary constructions is computing power—a difference that makes microgrids an even more compelling solution to grid calcification than they were before. A microgrid might be "an integrated energy system network consisting of distributed energy resources and multiple electrical loads and/or meters" (this is one official definition), but what is more important is that, despite being privately owned, they have a sort of republican spirit. Our grid could just as well be an amalgamation of ten thousand microgrids as a single system. So long as microgrids can function interoperably with each other, and do so most of the time, they are indistinguishable from the end user's point of view from the grid we already have. The difference, "resiliency," rears its redeemer's head only in moments of crisis when a microgrid's capacity to operate autonomously from the big grid works to keep the lights, pumps, and power on.

More like smaller versions of public power than instances of private power per se, microgrids are thus very different from what the Straws and their neighbors were building for themselves in the wake of the Great Gale. Not nanogrids (home-sized electrical systems that allow one to effectively live off the big grid), microgrids always have multiple sources of generation. These work to diminish the fragility of their supply chain, whether from oil embargoes or long spates of unexpected cloud cover. They also by definition have multiple customers—or meters—drawing, ideally, a diverse load from this little grid. Because their customer base and generation base just aren't as diverse (and can't be) as those that people and power the big grids, microgrids also need effective means for storing electricity. They also help to bolster the power storage market. For now, this means that most microgrids include large banks of batteries as well as some way to control load on the "demand side" (so-called controllable or sheddable loads). This usually means the ability to remotely shut down nonessential power-consuming devices like air conditioners or clothes dryers.

New Jersey Transit, for example, is building a microgrid for its commuter rail network, the first traction company to be making its own power since the days of Samuel Insull. This is a public infrastructure project, but it is also an electricity system with a sole owner. It's pretty big, covering a lot of ground and running a lot of trains, switchyards, stations, ticket machines, and offices. It can even be used to charge cars and phones, while creating safe houses—little bastions of heat and light—in its stations during hard weather. Despite the scope of the project, and its geographic reach, it's essentially a privatized power network. In this respect it is no different from a nineteenth-century street-car company.

Microgrids are their own special kind of thing—a new niche carved out between a system that relies solely upon "machines rotating in exact synchrony across half a continent" and an individual home that makes power only for itself. Microgrids are sharing machines that are not debilitated by that capacity when sharing means being sucked dry. They are not mechanical martyrs but self-interested, social, individuated electrical systems.

Though one can run a microgrid on 100 percent stock (nonrenewable) fuels, most also embrace a modest amount of renewable generation. NYU's is powered by a natural-gas-fueled combined heat and power (cogeneration) plant. The University of California, San Diego has added 3 MW of solar and 3 MW from a natural-gas-powered fuel cell to their 30 MW combined heat and power plant (also natural-gas-powered). Together, these warm and cool 450 buildings and provide hot water to 45,000 people. This system alone generates 92 percent of the campus's energy and saves them $8 million a year in electricity bills. Microgrids, as these examples make clear, often sidestep, ever so slightly, the green energy revolution hoped for by the Lovinses. Despite this, they also often end up being somewhat "light green" by accident. Because of the need for resiliency—which is to say, many ways of making power, many ways of delivering it, and many ways of using it, all of which can be adjusted in a pinch—almost all microgrids incorporate some form of renewable generation.

Despite the emphasis on multiple forms of power generation and multiple users (or, at the very least, multiple meters), these littlish, publicish grids, especially those built into urban areas, are designed to remain connected to the big grid most of the time. It is only in moments of duress that they disconnect and run on their own. The rest of the time, their flexibly produced electricity is available to all the grid's customers, and their load can be counted upon for balancing the big grid's production with consumption.

Some microgrids are necessarily isolated from the big grid all the time—most especially those built for isolated communities in the Canadian north or for mobile command units of the U.S. military in the Middle East. These grids are always an island. What makes these remote situations different from the kinds of microgrids being built for urban subdivisons or Sandy-besotted transit companies is that "resiliency" in these cases is not linked to their capacity to keep the lights on, even under the most extreme forms of stress. Rather, it means, more simply, that they no longer rely solely upon diesel-powered generators. The definitional diversity of microgrids means that they offer rural outposts a more robust and cheaper means for making electricity that have what the Lovinses would call the tail of a Manx—naturally quite short.

By eliminating or significantly reducing the amount of oil needed to run an isolated grid, of whatever size, supply chains, with all their attendant vulnerabilities, are drastically curtailed. More than 80 percent of the supply convoys in Afghanistan are, according to the Pew Charitable Trusts, "for transporting fuel and they repeatedly come under attack. The demand for electricity generation also weighs down our fighting forces and the rising cost of energy puts a major strain on military budgets." This concern is echoed by Peter Byck, who points out that "at three hundred thousand barrels a day the Department of Defense is one of the biggest users of oil. Getting that fuel to the front lines is incredibly expensive—in dollars but more importantly in blood."

Military bases until recently were 100 percent reliant on diesel generators. Not just for their lights, but their computers, and their air-conditioning, while the convoys that supply all this fuel are obvious and appealing targets. The military's most immediate goal, according to Phillip Jenkins, is to cut the amount of fuel it takes to support a marine in the field in half—from eight gallons a day to four.

Anything that can be done to eliminate the necessity of diesel generators, and reduce the amount of oil necessary to feed them on the field of battle, strengthens—adds resiliency, flexibility, and mobility to—the war effort. Mobile, matte, lightweight, and diversified systems for keeping the lights on, the data safe, and the troops cool are critical to mission success. For while some of this fuel is poured into gas tanks, a lot of it is used to make electricity.

Microgrids can, thus, provide grid resiliency in two quite different ways: at home they can help keep the lights on during severe weather or other "complexity" generated blackouts (like the East Coast blackout in 2003) and further afield they simplify supply chains that are inherently vulnerable to disruption or attack. As such, they add a degree of flex to two otherwise overly fragile systems. If what we need is resilient and reliable electrical infrastructure, then it seems our grids will need to get smaller, their interconnections will need to become more robust, and their fuel sources more diversified. Small is, quite suddenly, not only beautiful, it's also reliable.

If the rest of America is just now figuring out that grid-provided electricity is not always as reliable as we need, the military has known this for the better part of the twenty-first century. In part this awareness has arisen because of the supply-chain problem mentioned above associated with fueling generators, and in part because the load carried by a "dismounted warfighter" now ranges from 65 to 95 pounds, almost half of which is either portable electronic devices or the batteries needed to run them. On a four-day mission, both these numbers rise. The pack goes up to 150 pounds, and the batteries alone constitute a third of this weight.

It is simply impossible for our troops to be quick with over hundred pounds of gear strapped to their bodies. This is especially relevant today as war itself has changed since 9/11. American soldiers now find themselves in constant confrontation with a quick, light, flexible enemy that they can engage only in a plodding, obvious, and therefore inherently vulnerable way. The weight of the technology meant to ease victory on the field of battle now structures and limits our strategic options, and rarely in ways that make our soldiers safer.

In addition to being heavy, all of this technology is also wasteful and expensive. According to Theodore Motyka, a research engineer at the Savannah River National Laboratory, more than 80 percent of the energy needed to power devices like computer displays, infrared sights, global positioning systems, night vision, and other sensor technologies each soldier carries comes from *disposable* batteries. A brigade "will consume as much as seven tons of batteries in a 72-hour mission at a cost of $700,000." Just like the oil sent in coveys through the pass, these batteries have to be shipped in and later disposed of—a process that if not done correctly is incredibly toxic. A base now has its latrines to deal with (more on this below) and its depleted batteries. Most important, however, is that the transportation of liquid fuel into Afghanistan is so dangerous that the military prefers to fly in thousands of pounds of new disposable batteries than to truck in the gasoline that would be necessary to run the generators to rejuvenate rechargeables already in place on the ground.

According to Amory Lovins, at a typical forward operating base that was recently examined, 95 percent of the base's electricity went to "air condition, inefficiently, tents sitting in a hot sandy place." All of this electricity was produced by a diesel generator. He continues: "We are getting people blown up in fuel convoys to deliver the fuel to be wasted in that way. Just connect the dots and obviously there is something wrong with this picture."

At home the consequences are less extreme but the situation is not much better. Here, the most common, expensive, and disruptive forms of power outage are not the big storm-blown blackouts, but those of five minutes or less. These are rampant. Two thirds of the annual cost of outages in the United States are caused by those lasting less than five minutes, because of "the high frequency of momentary outages relative to sustained outages." Lots of little outages are disastrous for any industry that needs constant access to information networks and for which electricity maintains security, including electric door locks, key pads, metal detectors, surveillance cameras, and so on.

The military is both these things: they need to be able to collect and act on information instantaneously always. And almost all the means they use to collect and control access to this information, as well as to the spaces and machines used to contain it, rely upon electricity to work. Think of how many movies you have seen in which all the action is premised upon someone rendering a bunch of surveillance equipment inoperable for less than five minutes. This is not a scenario the Department of Defense ever wants to see happen under its watch, neither at home nor in the field.

Though the issues facing military installations and personnel might at first glance seem different—supply chains too long, batteries too heavy, existing electricity networks too brittle—in many ways the solution has been the same. The military too is turning to microgrids, of various sizes and structures, depending upon the specific needs of any given installation. It is estimated that in the next five years, the U.S. military will have added twenty-one microgrids to the twenty they already operate in the United States, more than doubling their electricity production to 578 megawatts. Many of these rely, at least in part, on renewable sources of power, as well as conservation and efficiency measures, which are lumped together under the term "demand reduction." For example, the army's tents in the Middle East are now covered with a thick layer of orange spray foam, which reduces the electricity needed for air-conditioning, heating, and ventilation by 40 to 75 percent. At Camp Lemonier in Djibouti, foam insulation was applied to the gymnasium tent. This alone reduced the number of generators needed to power the camp by 40 percent.

But "eskimoing" the tents in the desert (as this process is called) is just the tip of the conservation iceberg. Reliance upon both generators and portable batteries needs to be drastically reduced. Or, as a 2007 DoD report modestly put it: "There is a disconnect between consumption practices and strategic operational goals related to the security environment." To transform this disconnect into an alignment between strategic goals, the environment, and consumption practices, the Department of Defense has proposed a set of adaptations that bear much in common with the soft path. They are aiming for a mix of generation, storage, and efficiency measures that include, but are not wholly reliant upon, renewables or, for that matter, diesel; that are flexible, which is to say, they can be adapted to the particular needs of a base, or a reconnaissance team, or a brigade on foot; that are well suited to end use in terms of scale and of power quality; and that are relatively easy to understand. This last point is particularly important, for as one marine said to me in passing, during a discussion about the problem of developing solar "panels" for use in the Middle East (they can't be black, or shiny, or too hot, or rigid): "You give a marine two titanium balls in the desert, he'll lose one and break the other." If you are going to build a multipart, modular microgrid, to be assembled, disassembled, and cared for by soldiers, it needs to be hard to lose essential bits of it and even harder to break.

The solutions being implemented by the military are thus a surprising blend of technologies. For example, the Tactical Alternating Current System (or TACS) includes:

- 2.8–3.1 kW solar array
- power management center
- battery bank
- 4 kW AC inverter
- 4 kW backup generator

The main source of power generation is a "clean, silent . . . light-weight flexible and durable solar array. Excess power from the solar array is stored in a battery bank for nighttime and cloudy day use and backup power is supplied by a generator connected to the system's control unit." It should be noted that except for the generator used for backup power, this system is identical to the one Xcel gave the Petersons in chapter 6; the only difference is that for the Petersons, the grid provided backup power, whereas a generator does this for the military, isolated as it is on the plains of eastern Afghanistan.

In sunny areas, this backup generator is on only 5 to 10 percent of the time. When it's overcast it needs to run 15 to 20 percent of the time. Regardless this means that 80 to 85 percent of daytime electricity production is silent and emits no easily visible plumes of black exhaust.

According to the military's own calculations the TACS system also has 10 percent of the fuel and maintenance requirements of a diesel generator alone, and excepting the generator, it has no moving parts, which greatly reduces the possibility of losing or breaking any piece of it. TACS supplies 6 amps of continuous power at 120 volts, "the average load of a typical tactical operations center" with a peak load of capability of 60 amps, or ten times, the average load.

For larger applications, the DoD is also field-testing a "mobile power station" called, in high-acronymic fashion, the THEPS or "Transportable Hybrid Electric Power Station." Much like the TACS this combines "rigid solar panels, a wind turbine, storage batteries, and an augmenting diesel generator to guarantee continuous power during prolonged periods when wind or solar alone do not meet power requirements."

But here is where it gets interesting.

"THEPS provides, on average, 5kW of power output depending on the weather conditions. The inclusion of the diesel generator means the warfighter is not entirely freed from fuel logistics; however, even this challenge can be overcome if a system such as THEPS can obtain its diesel fuel via an in-situ resource such as biomass conversion." In other words, kitchen garbage and latrine sludge, burbling away in a specially designed tank about the size of a boxcar, generating biogas (farts and moonshine, or methane and ethanol) that can be siphoned off and used to run the generator and, not incidentally, power the cookstoves. They have called this the TGER (Tactical Garbage to Energy Refinery) a nicely self-contained digestion machine that the army has spent three years and $850,000 developing.

The day the military starts using their shit to make electricity, the world will officially have changed. It's an easy system; we could build it into our houses in about the same way we build a septic tank. Kitchen waste and old diapers (in the home setting), or toilet water and latrine gunk in the field, go into a receptacle and gas comes out.

The four to six pounds of trash a soldier produces daily could thus be "metabolized" rather than just burned. You do still need diesel, or biodiesel, to mix at a ratio of 1 to 9 in order to use TGER-produced gas to run an electric generator, but this means that with a THEPS plus TGER the amount of liquid fuel that needs to be trucked in to make electricity is roughly 1 percent of what is needed with a conventional generator-only system. This reduction has the ancillary effect of reducing fuel used not only for electricity production but for running vehicles as well. One estimate is that 70 percent of the gasoline actually used in the field is for transporting other gasoline around. Tanker trucks have to be fueled the same as any other vehicle.

All of these systems are still in development or being field-tested, but military-run microgrids on home soil are far further along, in part because these can rely on existing grid-provided power most of the time. Unlike a forward operating base, a military installation outside Hanover, New Jersey, or Washington, D.C., doesn't, for the most part, need to make its own power, compost its own excrement, or burn its own kitchen waste. It benefits from existing American infrastructures like flush toilets, utility-provided power, and garbage pickup just as much as do the average nonmilitary citizens of Hanover or D.C. These bases also suffer the same breakdowns of infrastructure when storms like Sandy take out or render dysfunctional our systems-in-common.

Far from existing in a situation of "energy poverty," which characterizes much of the world, including the war-embattled portions of the Middle East, domestic bases, much like the towns and cities that surround them, occupy a space of abundance. There is plenty of power and a good enough infrastructure to ensure that most of the time electricity gets to where it's needed. Nor are the infrastructural underpinnings of abundance limited to the grid. They stretch backward, first to the power plants and from there to the train cars and tanker trucks that transport the coal and natural gas we combust to make that power. Indeed they stretch further still, to West Virginia miners pulling down mountains to get at that coal and to circled semis around Wyoming bore holes extracting that gas. A secure, ample energy system also includes good employment policies that lead to peaceful labor relations and good county, state, and federal regulations that ensure both the affordability and the long-term stability of extractive technologies. And it includes making sure that disasters like Love Canal or Deepwater Horizon are a thing of the past.

Many of these integrated aspects of energy security can be controlled within the United States in ways that are impossible in war zones. The mistake is in imagining that the knowledge and organization necessary for abundance is naturally linked to energy security, or that a national infrastructure for supplying power to American towns and bases is situated in a "stable environment." We have seen this with Hurricanes Sandy and Irene in the Northeast and Katrina and Rita along the Gulf Coast, the Gale in the Pacific Northwest, Boston's Snowmageddon, the California droughts and their wildfires, and the annual spin of tornadoes that wipe out, and cause to be rebuilt year in and year out, the infrastructures that support northern Texas, Oklahoma, Kansas, and Nebraska.

But storms are far from the only way to make an unstable environment in the midst of an otherwise fairly stable nation. In addition to the headline-producing means of smashing through infrastructures and their supply lines, there are the things hardly spoken of, yet ever present. If the blackouts we remember as events of national importance are most often storm-made or complexity-made, there are hundreds every year, most in less urban areas of the country, caused by things that city people can't quite imagine. Like the time in rural Clatsop County, Oregon, when some guy got drunk and careened his car off the main drive and into the local substation (this was back in the day, before the utility put a big chain-link fence around it!), knocking out power to the surrounding area for two full days.

A similar, more malicious action was taken against the substation that supplies much of Silicon Valley with its electricity when, in April 2013, a vanload of masked men with submachine guns shot the rural installation to bits, and not at random. They targeted key bits of the substation's mechanics, like snipers against infrastructure, shot by shot destroying seventeen of the substation's large transformers. These men were never caught, but their know-how and malice made an impression. Though nobody talks about it, our physical infrastructure is not just exposed to weird weather, it's also shockingly vulnerable to weird people.

In 2014 there were seventy-seven serious power outages reported in the United States due to severe weather, another seventeen due to fuel shortages (usually the result of supply-chain problems, like congestion on the rails), and sixty-six that were the result of physical attack, only two of which were cyber attacks—a new problem that follows from computerized infrastructure. This count includes purposeful action, like the Silicon Valley substation snipers, and acts of stupidity, like the drunken substation smasher or the overly excited New Year's reveler (in the Straws' hometown) who loaded up his shotgun and, at the stroke of midnight, shot it exuberantly out his window. His aim was, unfortunately, a little low, and his blast hit the transformer across the street square on, causing it to explode in a fountain of sparks. The incident was a total accident, he wasn't in the least bit aiming at it. Or so the papers reported once the power was back up on that side of town two days later and the presses could run once again.

Then there is the constant roll of outages caused by plants (see chapter 5) and animals. According to one lineman in Santa Cruz County with whom I spoke, the main cause of residential outages was the incessant gnawing of squirrels, followed closely by limb loss in the county's redwood forests. According to a reporter for the *New York Times* who became interested in the actual impact of squirrels on grid functionality after learning that they had shut down the NASDAQ not once but twice (in 1987 and 1994), found "50 power outages" large enough to be worth reporting on were "caused by squirrels in 24 states" in just three months of 2013. In "two particularly busy days" in June, he reports, "Fifteen hundred customers lost power in Mason City, Iowa; 1,500 customers in Roanoke, Va.; 5,000 customers in Clackamas County, Ore.; and 10,000 customers in Wichita, Kan," all credited to squirrels. "In Austin, Tex., squirrels have been blamed for 300 power outages a year. Other utility companies have claimed that between 7 and 20 percent of all outages are caused by some sort of wild animal, and a 2005 study by the State of California estimated, hazily, that these incidents cost California's economy between \$32 million and \$317 million a year. Feral cats, raccoons and birds are also nuisances. Last month, reports surfaced in Oklahoma of a great horned owl dropping snakes onto utility poles, thereby causing frequent power outages." In 2011, a stretch of high-voltage lines outside Missoula, Montana, were shorted out by the carcass of a deer freshly killed by a juvenile bald eagle, who then picked it up and attempted to fly away only to discover (much to his chagrin) that a deer is too heavy for an eagle to carry. Down that deer came, accompanied by a hail of sparks, and once again, the power was out.

More recently and more endemically, there are the copper thieves. This problem further adds to the count of physical attacks against infrastructure. Times have been tough everywhere in America, and in the West and Midwest, where the space between communities is large and filled with quiet, the theft of copper grounding wire from substations has reached epidemic proportions. Sometimes the same substation is stripped of its copper several times in a month, the copper spirited away almost as fast as the utility can reinstall it. Transformers, which are wound with the metal, fall no less victim to the nimble fingers of copper thieves. Even utility lines—which I would be loath to use a pair of bolt cutters on, for obvious reasons—are not immune to being clipped and stolen for their copper. It even happens that giant spools of copper wire are lifted off the backs of utility trucks. Copper theft is an easy and profitable crime, if somewhat dangerous to the thieves and increasingly costly to customers and utilities alike.

Taken together the storms, the random drunkards with guns and/ or ill-steered cars, the carrion from on high, the incessant gnawing of squirrels, the rangy branches of trees, and the copper thieves have conspired for decades to leave people like the Straws and institutions like Google and the Department of Defense utterly unconvinced of the "stability" or "security" of their local energy environment. America may be blessed by a very real abundance characteristic of life in an advanced industrialized nation, but a first world lifestyle and energy security are not, it turns out, exactly the same thing.

What is curious about the new millennium is the admittedly gradual but ever more mainstream shift away from merely being disgruntled toward doing something about it. When speaking of the far left coast it may be easy for the dubious reader to attribute a back-woods spirit of do-it-yourself libertarianism to this switch from believing in and relying upon the grid (such as it is). This is not, however, what happened along the Gale-flattened Pacific Coast in the wake of the 2007 storm.

Not that there aren't a fair number of "those people"—the off-the-grid types who live in compounds rather than houses and hold radical political opinions considered far right or far left. But most of the folks we might traditionally refer to as "off-the-grid" have been industriously producing their own power, light, and heat since the late 1960s, mostly by the insecure (if we listen to Straw's and the DoD's sound opinion on the matter) means of diesel generators, though one does find the occasional waterwheel converting the unending rain into a tidy stream of volts. Small wind has never really caught on here, and solar, for those of you who have never had cause to visit the Northwest Coast of the United States, is not really an option.

There is a reason why the vampires in *Twilight* have chosen to live in Forks, Washington, and it is not the stuff of legend. The sun never shines. Sylvie Straw's small costal town 180 miles south of Forks has fewer than thirty cloudless days a year. Not only is it the most humid town in the nation (which is a nice way of saying that it rains all the time), but it was once, long ago, also voted the least good place in the United States to install solar panels.

What this all means is that off-the-grid, then as now, is not the same thing as being ideologically "green." Most people who produce their own power aren't doing it because they feel deeply that the electricity they get from the national grid is environmentally pernicious, though some do. Rather, absenting themselves from government services— and thus also, ostensibly, from the government's meddling—means producing power for themselves. Diesel gets you off the grid and so, too, does natural gas (an infrastructural network of another kind). To a certain extent, wood-burning stoves also accomplish this. But none of these embody the values that a distant observer might expect the bearded men and unshaven women of the region to subscribe to.

The Great Gale of 2007 did not bring about the awaited Left Coast, ecotopic, green revolution, nor did it help to affirm a Libertarian do-it-yourself-and-bugger-the-government's-meddling agenda.

It was rather that mild-mannered townsfolk, people with no alternative stance, people who would rather the government and the utilities did their job, had simply had enough. They began to take the project of heating and lighting their homes into their own hands.

When I say the world changed in December 2007, I mean that nonradicals began to take what would have been, even a year earlier, considered radical action. Townsfolk turned away from utility-provided electricity and invested their own money (these were still the boom years) to make something better. They did so not because of some strongly held antiestablishmentarianist stance but because they were tired of being cold and soggy, because they dreamed of a less hardy future. They dreamed what in a different time and place we might have called "immigrant dreams" but instead of imagining that somewhere far away across an ocean there is a better, warmer, brighter place, they looked at their own sorry, sodden lives and said: "Here. This place is my home. And it is in this place that I will build a better life." And so they have.

In this the Straws are not alone. Nor are the U.S. military or the new post-Sandy Northeast their only companions. Nice people, middle-of-the-road people, are building resiliency into the necessary infrastructures of daily life, not because they are opposed to grid-provided electricity on ideological grounds but simply because they are sick of its not working well enough to meet even the most modest expectations. By the time Sandy hit the greater New York Metropolitan area in 2012, the spirit of a bottom-up resiliency was already thick in the nation's air, even if on the East Coast they didn't know it yet. All they needed was their own storm to help them turn, faster than one might imagine, away from previous models of hardening infrastructure toward more contemporary plans for adding resiliency to it.

Even though the everyday actions of average people tend to get the least fanfare and the least public exposure, in the case of our grid these are precisely the activities that have set the tone for the revolution this infrastructure is currently undergoing at its edges. This is not a revolution in opposition to current structures of power, but rather a revolution that moves in concert with them. The great benefit of microgrids, whether one is Google or Fort Carson or Sylvie Straw's neighbor Bob, is that they work for people and institutions that are happy to use grid-provided electricity when it is there while providing them an option for when it's not; the responsibility for staying warm, well lit, and connected falls to them. Bob is just one man, not big enough or wealthy enough to have a microgrid, not even the scaled-down version suited for one house—a nanogrid. Bob lives in a modest apartment, he has a natural-gas-fueled heater and water heater. A propane-fueled stove (for coffee) and that shower. The Straws have much the same setup, though they have included a rotary phone, which doesn't need electricity to function, and propane lights. Many of their friends, the ones who care more about electricity and computing power and less about the temperature of their coffee and home, have opted for diesel generators. No one has yet conspired with their neighbors to build a TACS-style system for the block. And the city forbids small wind, for now.

These activities, the tinkering as much as more formal constructions, are collectively known as grid edge, a term that encompasses everything from hooking up a generator to a house or adding some solar panels to the garage roof or using a 50-MW acronymically named S.P.I.D.E.R.S. microgrid for your base, to building a wind farm, a substation, and private lines into your corporate headquarters. It is here, at the thick black border between being on-grid (and thus reliant upon an infrastructure that belonged solely to the electric company) and being off-grid (and thus having no power, as my own father did during the 1970s), that alternative visions and structures for power production are taking root. It is here, at the grid's edge, that microgrids blossom and grow. It is here that unstable environments are reworked into stable ones and that brittle is being remade in the idiom of resiliency. In the case of the military, this means standardized electric power regardless of the circumstances. For the New Jersey Transit Corporation it means keeping things and people moving even in the wake of an epochal storm. For people like the Straws it means hot coffee, hot water, light, and a phone without worrying if electricity makes this light, heat, and coffee or if it's achieved by other means.

Thus have Sylvie Straw, retired Brigadier General Anderson (champion of spray foam), Amory and Hunter Lovins, the New Jersey Transit Corporation, and Google all arrived on the same page. With very little fanfare, they have all stopped expecting the state and the utilities to do their job wholly by themselves. Taking matters into their own hands, they began—one by one, family by family, company by company, institution by institution—to figure out how to stay warm and dry and well lit, and how to keep their computers running steady and cool, when the next storm hits, or the next wave of Taliban mountain fighters hits, or the next complexity-generated blackout hits. They want to be able to do this without having to rely upon the big grid, which has at long last become too rickety and fickle a system for the constant production and distribution of electricity.

Sylvie Straw told me, in a follow-up conversation in January 2015, about how she and her husband (both now nearing seventy) had just gone to a Christmas party at the local senior center during "a big wind" and how the power had gone out just as they were all sitting down to eat.

"It didn't matter though," she said, because the dinner had been catered and "they were using that Sterno stuff" to keep the food warm. So they had a good hot meal with the rest of the local aging folks. Even the raffle was run by candlelight. The Straws won a twenty-five-dollar gift certificate to a local Finnish-themed store. "Daniel," she told me, speaking of her husband, "always wins at raffles." It was the fifth outage already that winter, which was only two months old.

Despite their individual efforts to build energy security into their home lives, the Straws are still, in 2015, living their lives at the far edge of the grid's capacity to provide reliable electric power. They'd be the first people to cheer a microgrid for their town, but until that time, they will suffer what their common infrastructure fails to provide while nesting more or less comfortably in what their individual efforts have accomplished.

CHAPTER 8

In Search of the Holy Grail

Toward the end of 2014 Lockheed Martin's super-secret research and development wing "Skunk Works" made a surprising announcement. Even more surprising was the brouhaha with which it was greeted in energy circles and even some mainstream media outlets. Their claim was to have shrunk the time horizon for a viable fusion reactor from thirty years to ten. This modest change in the time-until-release of a nonexistent technology would seem a funny thing to have made such a big splash. To understand why it did, it is necessary to know the backstory, now transformed into a running joke, about the future of fusion.

For decades it has been said that nuclear fusion is thirty years away. In the 1950s we could be certain of its arrival in the 1980s; by the 1970s, the year 2000 was the promised time; in 2010 it seemed that the kinks might be worked out of the process by 2040. Before Lockheed Martin stepped in and shrunk this ever-receding future down to ten years it seemed that our cars would fly, our refrigerators think, and nanobots swarm through our bodies eating our cancers before nuclear fusion would ever power our world.

Despite the retreating horizon of its success, fusion has, at least since the 1950s, been the promised end point of our energy woes. As a "clean" energy source, fusion produces no radiation, it uses a neutral isotope found in water as its fuel, and the reactor that produces it can't melt down. With fusion we would have a limitless source of nonpolluting power. If Lockheed Martin wins the race toward a fusion-powered future we will even get a smallish reactor that can be both mass-produced and strategically deployed. Imagine the tiny town of Anaktuvuk Pass, Alaska, with a fusion-powered microgrid. No supply chain, no pollution, total energy, total energy security. The problem is that it takes about as much electricity to run a fusion reactor as that reactor produces. It is a zero-net-energy machine; the power that goes in equals the power it spits out. Nevertheless, for over half a century, fusion has been the electricity industry's holy grail: limitless power from water.

Fusion hasn't lost all its sparkle; it still maintains a tiny glint of grailness. The French have been building a very big fusion reactor for the better part of the last decade, and Lockheed's efforts to come up with a smaller machine also prove that fusion-fueled dreams still have the power to motivate substantial research and development projects. Upon reading the small print of Lockheed's announcement it becomes clear, however, that the company's ten years might as well be thirty. The press release was basically a job ad. Lockheed does have a fusion reactor on paper and is casting about for "fusion experts" to help translate the thing from design to prototype. And, as they search for the right guys to make their sketches into reality, other holy grails rise and fall from favor. So is it today, so has it always been. What counts as the holy grail changes as our visions for a more perfect future are themselves transformed by present-day concerns. Today the grail is less a new way to make power than it is to find a really good way to store it.

Engineering a means to effectively store electricity is not a new problem. It's been a priority for those involved in making, and making money off of, electricity since the Insull days. Figuring out this problem has risen up to grail-level urgency today because every dream for a cleaner energy future that involves a lot of renewables requires that we have some way to put aside the too much electricity they are capable of making.

At present, our capacity to integrate renewable generation gets complicated as we approach 15 percent of peak power—or 25 percent of daily electricity use. This does not change with the size of a grid. A diesel generator is a better power system than a single solar panel, unless that panel comes with a battery pack. A 500-megawatt coal-burning plant is a better power source than are 500 megawatts made by a wind farm, unless a means of balancing the wind is built into the grid that that farm powers. And though we do have batteries robust enough to back up a home solar system, for the moment most people installing these systems don't bother with the expense. After all, they have the grid to help them out when their panels fail. Scale up the renewables and storage becomes even scarcer.

For the moment we solve this problem by doing things backward. Rather than storing excess power we use generation to "balance" generation. Every time a cloud goes by and diminishes solar output for a second or two, we burn some fossil fuels to generate enough little jolts of electricity to even out the electron flow. If we use a traditional power plant for this job, it will operate at only 2 percent of its productive capacity. These massive machines were built to be efficient only if running full bore all the time. With the introduction and spread of renewables we now require something else, something not as staid as a big coal-burning factory and not as mercurial as a wind turbine, something that can run for a little while on call, be it two or three seconds, or two or three hours (but rarely two or three days).

Up until now the most common means for doing this has been natural-gas turbines, which most large renewable-development companies also invest in as they build their wind farms or solar fields. But nobody says this "something" needs to be a new form of electrical generation, though if it is we'll need to learn to think more symphonically.

"In a symphony," says Clay Stranger, "no instrument plays all the time, but the ensemble continuously produces beautiful music." So too, he suggests, might we learn to conduct our grid, more artfully and with less of a lead foot. Storage at this point is a far more popular solution and a lot of people are working on it with some admirably creative ideas entering the mix. There are also plenty of slightly boring but efficacious ideas floating around. What all of them have in common is a gut-felt understanding that a way to store electricity is necessary, if we are to meet the future with open arms and an optimist's heart.

As improbable as it may seem, though we have been making and using electricity for nearly 150 years, there is still no way to put it aside for later use. One can't loan a cup of watts to a neighbor who is short a few to bake a cake. One can't fill barrels with the stuff and then load these onto train cars or ocean-going tankers and ship them across continents or overseas. Imagine, if some of that way-too-much wind power from the Columbia River Gorge could be packaged up and sent by rail or interstate to West Virginia where coal still fuels 98 percent of the local power plants. There it could sit in a warehouse until it was needed and then this nicely aged green power could be emptied into the local substation and sent on its way. It could be a year, or thirty years, later, and the electricity would be just as fresh as when it was plucked from the air. As odd as it sounds, this is precisely how oil and coal work today. It doesn't matter when or where they were extracted, and it doesn't matter how long it takes for us to get around to using them. They can wait. Time, the crippling, confounding factor in electricity production, is almost not a factor at all when thinking about the other things we use to fuel our world.

Most of us, when we think about storage, think first of batteries, but though batteries may be good for bringing electricity with you, they have not, until recently, been very good for storing power at grid scale. At the moment there is only one battery on our grid: a 22,000-square-foot, 1,300-ton nickel-cadmium battery that was built outside Fairbanks in 2003. This can supply 40 MW of power for about seven minutes, though it is most often used for twice as long at half the power. Though it does keep the lights on and heaters running for a bit, it isn't exactly a backup-power system for the grid; it's a stopgap that allows the local utility the necessary quarter of an hour that it takes to fire up their diesel generators. These then run the grid for the duration of the blackout. For the moment, this is both the biggest chemical means for storing electricity at "grid scale" and also the one with the longest duration. The rest of the meager means we do have for storing electric power for our grid are electrically driven mechanical processes that can be reversed to regenerate an electric current and these options are all limited by topography rather than technology.

For places with mountains, there is "pumped hydro." It's a sort of man-made dry lakebed near an existing hydroelectric dam. When there is too much water, whether run-off or rain, some of the electricity the dam makes is used to pump the excess water uphill, into a second reservoir, where it sits until additional power is needed, then this water is allowed to flow naturally downhill again passing through a set of turbines at the bottom to generate a "new" electric current.

Ninety-five percent of the electricity "stored" in the United States is guarded in this way—22 gigawatts' worth or the equivalent of 2 percent of our national generating capacity. Pumped hydro works great in places with hills and dammed rivers. It's less great in the prairie, swamp, or desert states. From Nevada all the way to Indiana, there are effectively none. And where the South flattens out, so, too, do the pumped hydro stations thin and disappear.

A couple of Gulf states do have something as useful for electricity storage as a hill near an existing reservoir. Alabama and Mississippi have salt domes. Starting just north of Mobile and stretching all the way under southern Mississippi are a warren of natural salt caverns long used for dumping toxic chemical waste. These can be just as profitably used to store compressed air. This is precisely what the CAES plant, in McIntosh, Alabama, does. When electricity is cheap or there is too much of it, usually at night, the excess is used to condense air and force it into these caverns. Then, during the day, when demand is high, the air is released. As it expands, or decompresses, it spins a turbine to regenerate an electric current. Unlike pumped hydro, however, this air isn't storable indefinitely.

Compressed air is a twenty-four-hour affair. Electricity is used to "charge" the plant during off-peak hours, and decompressing air is used instead of coal to make electricity during peak demand the next day. Unlike a battery—the workings of which a compressed air plant mimics—the benefit of a mechanical rather than a chemical "charge-discharge" storage system is that it's good pretty much indefinitely. The Alabama plant has been running this diurnal cycle, day in and day out, for twenty-five years (it was completed in 1991) with no appreciable loss in efficacy. There are no batteries in existence that can compete with this, though "flow" batteries—a grail many feel is worth questing for—seem to hold the promise of almost twenty years of rechargeability. Flow batteries are, if rumors are correct, only a year or two in the future.

Another even more recent stab at energy storage has come in the form of concentrating solar towers. In the United States we have two of these, both in Southern California, with a third nearly complete in Nevada. Similar to pumped hydro and compressed air, concentrating solar towers employ a battery-style logic. An array of mirrors is situated in a sunny place, usually a desert, with all of their angles adjusted such that their individual beams of sunlight are directed at a looming central tower filled with something rather like table salt, which liquefies at right around 530 degrees Fahrenheit.

The first two months after the plant is turned on, all the force of the sun is needed just to melt the salt; after that it remains liquid, its temperature rising and falling with the diurnal cycle. The sun's daytime heat is stored in the liquid salt until needed, and then that heat is used to boil water, which drives a normal steam turbine to generate electricity.

If pumped storage can hold "the potential to generate electricity" indefinitely, molten salt towers, like compressed air plants, work for only about twenty-four hours. Solar trough plants, some of which also have molten salt storage, are good for about six hours of power production after the sun sets. This may seem like a brief period, but in solar-powered states, six hours is all that is necessary to get everyone home from work, fed, TV'ed, and into bed. At this point electricity use plummets and molten salt becomes a little too cool to continue to do much good. There are currently eight solar trough plants in the United States. Six are in the Southwest, split unevenly between California, Arizona, and Nevada, and half of these, including the big ones (over 250 MW), have come online since 2013.

That's about it. Grid-scale storage, an element we need in order to integrate significant variable generation into the grid and to deal with our market proclivities for selling and shipping electricity as if it were a regular commodity, is limited to this: some artificial lakes, one compressed air plant, three molten salt towers, eight solar trough plants, and a lot of dreams about batteries. Each of the existing systems is a custom-designed one-off. We may be able to store electricity for all of southern Alabama, but we can't yet do it for the few solar facilities that make the meager 0.8 percent of that state's power. Yet we are going to need dispatchable electricity at a grand scale if we are to have any hope of even approaching the goals set during the 2015 Paris climate change talks and similar sorts national and international of conversations. These include: "Eliminate burning fuels wherever possible in favor of using electricity" and "Generate electricity with clean power sources—such as wind and solar—and eliminate coal- and gas-fired power plants." And we need dispatchability and probably storage now, at a smaller scale, so that the one technology that people actually seem to love—the solar panels—can be a part of the renewable-energy revolution we are more globally attempting to bring to pass.

People like to imagine this will be a battery but it needn't be. In fact, limiting our imagination in this way from the start is probably the worst thing we can do as we move into the twenty-first century. As the anthropologist Akhil Gupta reminds us: "We need to reimagine electricity use in the future that does not simply seek to extend patterns in the present." Rooftop solar forces our hand in this regard. The solar revolution is already well under way, and its particular failings (namely, solar's diurnal cycle, its minute-to-minute jitter, and its tendency to overload the grid at noontime) are pushing us right now toward a broader imagination of "other possible electric futures" for storage, for more efficient design of everything from house walls to clothes dryers, as much as how, where, for whom, and by whom electricity gets made.

Hawaii is neither the sunniest place in the United States nor the most cloudless. But it's sunny enough, and its electricity rates are high enough, that solar power generating capacity in the state has doubled every year since 2005. Hawaii now has the second-highest penetration of rooftop solar in the country (after Arizona) and the quickest payback on the investment anywhere. A home-owned rooftop solar system in Hawaii will pay for itself in just about four years—or faster than a car loan. Part of this is because Hawaii is the only state in the nation that still makes its electricity from oil, floated over on tankers from the mainland, and thus has electric rates more than twice that of any other state and almost three and a half times the national average. This extra-expensive electric power, in a mid-ocean archipelago where most things already cost more than in the contiguous states, is part of what has convinced Hawaiians to opt for rooftop solar. More than 12 percent of them now have panels on their roofs. And these

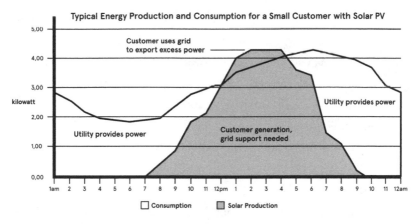

FIG 5 Solar production and electricity consumption vary over the course of a day. From noon to about five P.M. solar, in some states, can produce more than the electricity consumed by everyone plugged into the grid. Without a means of storing or using the excess, this bounty makes little impact on the needs of utility customers during all the other hours of the day. (Reproduced with permission from the Edison Foundation Institute for Electric Innovation, "Value of the Grid to DG Customers" IEE Issue Brief October 2013. http://www.edisonfoundation.net/iee/Documents/IEE_Valueof GridtoDGCustomers_Sept2013.pdf)

privately owned, rooftop-mounted solar systems on bright sunny days often produce more than 100 percent of the State's electricity needs.

This "more than 100 percent" of daytime demand is the number to keep in mind when people cite national statistics regarding solar power production. It is true that nationally 0.6 percent of the electricity on our grid is currently made from solar, but this aggregate hides local truths (and worse, in many places, it doesn't include the power made by home owners at all but only the solar from big, corporately owned plants). In certain densely populated sunny places, like Hawaii and Phoenix and likely soon the whole of the L.A. basin, privately owned solar feeding the grid provides up to and, at times, more than the daytime electricity needs of local people, factories, and businesses. As with wind power (at 5 percent nationally) local developments in small solar are more indicative of problems that still lurk on the horizon for most of America than they are of "national" realities, produced from aggregate statistics.

Though privately owned, all the electric power made by rooftop systems, everywhere in America (including Hawaii), is fed by law back into the public grid. These aren't private power systems; they are tiny public power plants that function, effectively, like a big power plant in shards, its many megawatts made in 10-kilowatt packages by any private citizen willing to foot the initial investment. Each roof a power plant, not for the house it sits upon, but for all of us. Their power feeds the grid every bit as much as a centralized coal-burning plant or a hydro-electric dam. And these small producers are using power from the grid too, at night, when there is no sun, on cloudy days and even minute to minute as the electricity their panels produce runs in little jagged peaks and troughs out of the panels and into our common set of wires.

Here is how it works. A home or business owner, considering the empty and non-profitable space of her roof, calls a panel company, which comes out and waves something rather like an iPad-on-a-stick around on top of the house and from this determines what portion of the roof would need to be covered in photovoltaic cells in order to offset actual electricity use. Then they offer a package that spreads the cost of these panels over a set period of time, in the neighborhood of 240 months. This rate per month is designed to zero out a panel owner's electricity bill, and it is usually about what a customer was paying to their utility to start with. The main difference is that like a mortgage this rate is invariable and, critically, it isn't paid to the utility. The money rather goes to the panel company or, in certain cases, to the bank. Most often what this package includes is the panels, installation costs, another meter, a DC/AC converter box, and maintenance. It could also include a big battery or some smart appliances, but for the moment it does not. Then, in theory and often also in practice, once the panels are up they feed roughly the same amount of electricity into the grid over the course of a year that the house or business pulls out of the grid over the same period.

The numbers might not cancel each other out day to day, what with clouds and weak winter sunlight, but over the long term the amount of power in and power out is structured to be more or less equal. The electric bill, always a tricky bit of documentation to understand and an impossible one to control, is thus nullified, and the only thing the home owner has to come up with at the end of the month is a known, mortgage-like payment to the panel company.

As it has been engineered by the panel providers and the electric company this decision is neither one of increasing energy security nor of greening home electricity consumption. Rather it comes down to dollars and cents. Will it be cheaper over the next quarter of a century to have solar panels or not to have them? The answer, quite often, is yes. If the angle of your roof is good and there aren't too many trees or tall buildings around, it will. Especially given the tax credits and rebate systems currently in place.

This turn toward economic rationality and away from ideological points of argumentation for solar has caused a lot of people in America who don't think much about the environment to adopt home solar systems. While those who do care about making the world a better place by making power differently were always an easy sell. Together, the two groups have contributed to a 1,500 percent jump in solar panel sales since 2009, and not just in the Southwest or on tropical islands. The Live Free or Die states in the Northeast are also proving to be rapid adopters of rooftop solar despite not having anything like the sun of their fairer-weather brethren. Even without perfect conditions, residents in New Hampshire, Vermont, and Massachusetts are betting that they can still have a smaller monthly energy bill with solar than without. Maybe not in the twenty-teens, but certainly over the long haul. Electric bills, after all, have a tendency to go up, while the monthly payment on the panels is guaranteed not to.

Another pocketbook-driven cause for the massive jump in the amount of solar in the United States in the past five years is that the price of very good panels has fallen by half. This is largely due to incentives to Chinese producers for the German and Japanese markets, but the United States has been happy to get in on the boom times. As the price of panels goes down, paying for grid-provided electricity becomes less and less appealing to home owners of moderate to substantial means, especially in places where big-grid electricity is expensive (such as Hawaii) and in civilizational pockets where consuming locally produced products is part of the local cultural ethos (such as Vermont or North Carolina).

Had more electric companies gotten into the solar panel business earlier, this switch from paying a utility bill to paying a panel company for electricity might not have led to the dire situation faced by our grid today. As more and more people opt for the panels, the electric company is left with little to no revenue with which to maintain the grid—the way the utilities have dealt with this in Hawaii, but also in Southern California and especially in Arizona, is to raise rates on the customers they still have. People without panels end up paying much higher utility bills to maintain the infrastructure that the people with panels still use every day (as they pump electricity into the grid) and every night (as they draw electricity down from it). They use the system, we all do, but grid maintenance is left on the shoulders of the remaining ratepayers. And though this isn't fair, utility attempts to get reasonable line access fees or standby fees (that allow solar panel owners to use grid-provided electricity when their systems fail) passed in states with high solar penetration have foundered. For years the utilities used these same rate structures, set at exorbitant levels, as well as fees of every imaginable sort—like demanding thousands of dollars just to read the applications of wannabe solar power producers—to keep home panel owners from entering the market at all. State legislatures thus don't trust the utilities not to take a mile if granted an inch, and they deny them any access to rates or fees that might be manipulated to reduce alternative power generation.

It used to be the case that the nonadopters of rooftop solar were renters, or people of such modest means that even if they did own their homes they did not have the resources to buy panels. These, the poorest and most transient members of our population, were stuck with their utility, and so they were also stuck footing the bill for the rest of us.

This loophole was soon closed, however, with solar panel leasing programs, which now account for 75 percent of the rooftop solar in Southern California and 85 percent in Arizona and with solid inroads in the Nevada and Utah markets. Now anyone without a mountain of debt and with a roof can rent the panels and still pay less than their normal monthly electric bill. For their part, community organizations are helping to fund opt-in installations for apartment dwellers and encouraging developers to build rental units with the panels already installed. The lease, or the loan, is then folded into the cost of renting the space.

The true owner of the panels is the company from which they are leased and they cover all installation, maintenance, and equipment costs in exchange for the right to collect state and federal renewable energy subsidies. They also receive payments from the customer over the life of the lease. Everybody wins. Except people with bad credit, unstable jobs, or transient lives. Except the utility. Except the grid.

The ever-shrinking number of grid-only-electricity customers are rightly irate. The cost of maintaining our grid in common, the grid we all still use twenty-four hours a day, seven days a week, fifty-two weeks a year, is now on their backs. Because rooftop solar installations don't take people off the grid, and they aren't transforming individual homes into islandable nanogrids or neighborhoods into discrete, sustainable microgrids. Indeed just the opposite. Distributed solar causes the once "somewhere" of electricity production to start to look a lot like the more familiar "everywhere" of electricity consumption. But while electricity is increasingly made everywhere and also used everywhere the grid still stands as it long has between production and consumption; it still interconnects us all.

The grid as a common infrastructure is what makes distributed solar possible. And, because we don't have a way to store any of the electricity these home systems are making, distributed solar is flooding the grid with electricity, at least for part of the day, while starving it of money.

All of this new activity is happening on the bit of the grid designed for distribution—the low-voltage wires slung between homes and pole tops in residential neighborhoods. Ironically, this is also the weakest part of the system, the most likely to give way, and the least well kept up. Between the 1950s and 1980s, outages increased modestly, from two to five significant outages each year, compared with 76 in 2007 and a whopping 307 in 2011. Almost all of these more recent blackouts, much like the big East Coast blackout of 2003, started and propagated on underfunded distribution networks.

Meanwhile transmission systems—those long high-voltage lines that stretch between distant power plants where electricity was once solely made and the urban centers where it is still mostly used—are much less prone to outages. This is in part because of their simplicity and their height. It's harder for a tree to fall on a line 110 feet in the air. Their relative reliability is also the result of substantial commitment to upkeep over the past fifteen years. Since 2001, investment in high-voltage transmission infrastructure has held steady at about 7 percent a year, while investment in low-voltage distribution networks, including smart meters, is below the necessary threshold for basic maintenance.

One analyst nicely summed up this industrywide care for long-distance wires when all the new activity, and the stressors that come with it, is happening on local, tightly woven, short distance wires, saying: "A continued reliance on centralized generation and the relative fragility of high-voltage transmission lines is completely out of alignment with the growing acknowledgment amongst regulators, political leaders, and industry that grid resiliency is not addressed in this way."

A recent special report in the *Economist* took it one step further. As a result of the veritable explosion in rooftop solar, "the power grid is becoming far more complicated. It increasingly involves sending power at low voltages over short distances, using flexible arrangements: the opposite of the traditional model."

The grid, in other words, is changing. We, the people, are changing it. The utilities, so long the only game in town, are not keeping up. It's not just a question of correct investment, it's also about thinking on their toes. They don't. As a result lighter, quicker competitors, some of which are companies, many of which are just regular folks, are running circles around the traditional utilities. Though the situation in the United States is dire, in many parts of Europe, and in Germany most especially, the crash that experts fear here has already happened. While we talk about the utility death spiral—where renewables ruin the grid by putting the utilities out of business—as something that might still be avoided, something that large-scale and small-scale electricity storage might stave off, in Western Europe it has already begun.

In 2013 Germany's two largest utilities lost a collective $6 billion as many of that country's corporate entities got off the grid altogether. In the United States we raise electricity prices on poor and transient populations (people who don't own houses, mostly); in Germany, however, they raise rates on businesses and manufactories. The companies are the ones being asked to foot the bill for the expansion of renewable generation across all sectors. As a result, they have begun to walk away from grid-provided electricity entirely. They can do this because unlike renters, German companies have the capital to build themselves private plants. If, in 2013, 16 percent of German companies were already energy self-sufficient, this was 50 percent more than the previous year, and another 23 percent were actively investing in a near future defection.

This is what is meant by the utility death spiral: "as grid maintenance costs go up and the capital cost of renewable energy moves down, more customers will be encouraged to leave the grid. In turn that pushes grid costs even higher for the remaining customers, who then have even more incentive to become self-sufficient."

Meanwhile, utilities are stuck with a bunch of stranded assets. Those big, expensive power plants the utilities built throughout the twentieth century aren't needed much, if at all, anymore. They are still being paid off, however, even if they are largely inactive. Much like all those unfinished nuclear cooling towers that dotted the American landscape in the early 1980s, big fossil fuel power plants, in Germany as in Hawaii or Arizona, stand as a testament to massive investment in the wrong path. The CEO of the German utility REW, who has been charged with overseeing his company's collapse, admitted that the utility invested too heavily in fossil fuel plants at a time when it should have been thinking about renewables. "I grant we have made mistakes," he said. "We were late entering into the renewables market—possibly too late." All of this, he continued, adds up to "the worst structural crisis in the history of energy supply."

If variable generation is bad for the grid, then distributed, renewable generation is worse. If utilities have been slow to adapt to a customer base at long last given a modicum of control over their bill, then government actions, mostly in the form of subsidies that support ever-increasing renewable generation goals, have transformed utilities into dinosaurs. In 2015 the utilities are lumbering remnants of a twentieth-century way of doing things. And though many are now scrambling to find new ways of generating revenue, they are hamstrung at every moment by a regulatory structure that impedes quick changes and trial balloons.

One thing is clear: American utility companies cannot maintain the transmission and distribution systems on our grid by charging customers solely for how much electricity they individually consume. Customers, homeowners as well as industrial concerns, want to use less power, want to make more power (and get paid for it), and want the grid to make these things possible. New modes of billing will need to be developed (and some are already in the works) but because utilities don't set their prices or develop their billing schemes themselves, all of this work must pass through a regulatory bureaucracy, and in some places also through a state's legislature. As such, easy fixes are all but impossible to come by. What is perhaps most surprising about all of this is the degree to which the conversation about solutions to the revenue problem are only just getting started. It is as if the utilities woke up one chilly morning in 2014 and realized that their ship was about to be sunk.

All of this would also be playing out differently if only there was a way for the utilities to put some of this locally made solar power and farm-made wind power aside for evenings and winters, calm and cloudy days. Finding a way to store electricity so that it is there to be used when we need it is at the top of every ideologue's list. This is true whether one wants to save the big grid or replace fossil fuels entirely with 100 percent renewable generation; whether one wants to chop the existing grid into overlapping, interlinking microgrids, or just create a universe of islands in which every enterprise has a private grid and every home is an isolate. While different visions of our collective future produce a need for different kinds and scales of storage, they all rely upon electric storage of one form or another. As a result, if one can look beyond the battery, the work being done on this front is both prodigious and intensely imaginative.

I saw a man in Stuttgart proposing "buried" hydro in place of "pumped" hydro storage for the Germans. Pumped hydro is impracticable in Germany because the citizenry like their mountain valleys for tourist activities and dwelling places. Instead, he suggested, a great mass of earth could be perforated all the way around, a bunch of giant O-rings installed under it, and then excess electricity be used to pump water down into the earth, raising this "plug" of land up to 275 feet high. Imagine, perfectly cylindrical artificial mountains rising up in the middle of fertile German plains. He'd done the math and he was serious. These pistons of bedrock would then be allowed to sink, with gravity, pushing the water out again and through a standard turbine to create electricity when needed, using essentially the same physics as regular pumped hydro storage.

Closer to home, there is a grand plan afoot in the West that would hook up compressed-air storage in Utah to wind farms in Wyoming and then, using an existing high-voltage line from the more southern installation, to send power to Los Angeles—a line that is becoming available because the last coal plant it was designed to serve will be retired in 2027. The line will be serviceable but empty; the salt caves, prepared and treated to hold natural gas, will also be empty (the shortages of that substance proved a fleeting fear), while the fierce bluster of wind across Wyoming will at long last be harvestable for an urban population dense enough to make use of it. This project shimmers with the future-possible. It is practicable, like the basalt land plug, but far further along in the permitting and funding process. It will gather its monies from a mix of private investors, government subsidies for kooky but viable ideas (called ARPA-E), and state and federal funds set aside for infrastructural upgrade.

The feasibility of these kinds of big, better-than-a-battery projects is not dependent upon scientific ingenuity so much as the difficult process of translating ideas through real-world bureaucratic and cultural systems, gathering both money and support along the way.

All too often the inventors of the next big thing are so focused on the fact that they have found the grail that they forget how hard it is to materialize and then sell it. A great many potential solutions to our grid's problems have been sketched out on company drawing boards, all of them capable of radically changing how we make, use, and even store the power we need in order to live and prosper. The only ones that matter, however, are those capable of moving from sketch to prototype, from prototype to extant, and from extant to ubiquitous. It turns out that the more land to be disrupted and the more visibly unusual that disruption, the harder it is to move even a brilliant project (like the basalt plugs) off the drawing board and into our common landscape.

To get from here, where we have an immediate problem of too much variable generation, to there, where we can use this "too much" to help us get off of fossil fuels entirely, will involve a lot of small steps in more or less the same direction. And because change, especially of this scale, makes people nervous, it turns out that the storage solutions winning the minuet toward the future are the ones that either don't look "infrastructural" at all (like repurposed salt caverns) or those that at the very least mimic the infrastructural forms of things we are already comfortable with, like how pumped hydro storage looks like, and is, just another reservoir. The closer new projects of infrastructural revolution come to more populous areas and the more they stand out, the more stringent resistance on all fronts becomes.

This invisibility, or capacity to disappear into the given, is also a part of what storage needs to accomplish in order to succeed, not physically so much as culturally. And mimicking familiar urban "skins" seems to hold a great deal of promise for the adoption of new electricity-storage technologies, both large and small. Three familiar forms are getting most of the attention from consumers and the press; these are the air conditioner, the office tower, and the car.

Let us begin with the little one. Air-conditioning, as we well know, is the grid's true nemesis. If the grid were a James Bond movie, the air conditioner would play the villain. A subtle madness in its eye as the planetary weather warms and as climate-controlled spaces become increasingly normal, air-conditioning bides its time. It is waiting for that moment when we all, of our own accord, fire it up and the need for electricity *now* shoots through the roof, the whole grid sagging, cracking under the weight of our collective demand. No true Bond villain could hope for a more nefarious outcome.

Everywhere, and often inefficient, air-conditioning is finally being replaced in some places by an ingenious, if old-school, energy-storage device—an icebox that uses electricity during the middle of the night to make ice and then blows hot daytime air over that same ice during peak demand hours. Where an electromechanical air conditioner, the kind most of us use, employs electricity to move hot air over a refrigerant to be cooled and dehumidified the icebox uses a fan to blow hot outside air over ice. By this means the air is "conditioned," as the ice melts and returns to water (ready to be refrozen the next night) and the edifice to which the unit is attached is kept at a comfortable temperature during even the hottest hours of the hottest days. All of this is accomplished using about the same amount of daytime electricity as a ceiling fan. It's an icebox and, effectively, also an electricity box—nighttime electricity stored in the form of ice.

Of all the recently introduced means for storing electricity on a small scale, the Ice Bear energy storage system, as it is called, is the only one that has really taken off. Homeowners buy it, subdivision developers invest in it, and manufactories, businesses, and even data centers mount it. This icebox, graceful in thought if familiarly clunky in form, is a winner.

Radical innovation, in wide deploy, ends here. All of the rest of the mainstream dreams, and the biggest investments in realizing these dreams in the twenty-teens, come back to the battery, though it is often quite cleverly disguised as other things. If in 2010, before distributed solar really took off, the storage field was wide open, filled with fantasies of giant flywheels; networked, smart hot water heaters; expandable hydrogen-filled balloons; and hyper-heat-absorbing ceramic tiles, the wonders of this brief early moment of grail dreaming is now radically circumscribed. Batteries were not so very long ago a minor part of a wide-ranging conversation about storage. They were not even in the top tier of not-very-good options because they were too unwieldy and expensive. The chemistry of the best of them was insanely toxic to manufacture and to dispose of, and many depended on rare earth elements found almost exclusively in China. The limited rechargeability of early twenty-first-century batteries and their precursors was simply not sufficient to merit the initial investment. All of this changed with the rise of practicable, affordable lithium-ion batteries, which have since 2010 taken over both the market and the imagination. A conversation about storage today is 85 percent a conversation about present-day and future battery technologies and 15 percent a conversation about weird ideas that somebody made work once, someplace suspect, like Alberta.

An interesting success story is a newly purchased, though not yet built, storage facility just south of L.A., in Long Beach, California, which when complete will house the world's largest electrochemical battery. Not that any casual passerby will notice, because this mega-battery is being built to look almost exactly like an office building. Parts of it will even function like an office building, but for the most part it will just be thousands upon thousands of stacked lithium-ion batteries, capable of generating up to 400 MW of electricity (though it can run at this rate for only four hours). And though it might look like an office building to us, from the grid's point of view it might as well be a normal gas-fired power station.

This project maintains one of the distinct advantages of batteries over the other kinds of mechanical, or even chemical, storage on the market: namely, it can be scaled up or down by the millisecond, useful for balancing out the power generated by solar and wind being pumped into the grid by all those rooftop panel owners and wind farm conglomerates. It fails, however, on another front. Like a salt cave, or a pumped hydro reservoir, this massive office tower is not a portable storage device. It's stuck where it is. As such, it fails to capitalize on the other distinct advantage of the battery. It's not just that they can be scaled, it's that they can be moved. A battery and a barrel of oil are not, in this regard, so very different. Even less so when they are (each in their own way) inserted under the hood of a car.

Batteries, despite their ability to produce electricity on call, don't actually have electricity inside them, instead they are full of chemicals. Under the right conditions these chemicals can be coaxed into a reaction that causes chemistry to produce electricity. In order to work, each of a battery's two "terminals" has to be made from a different kind of metal separated by an electrolyte. Any number of things can serve as an electrolyte, from soda pop or a potato to sulfuric acid or even ceramic, though various kinds of salts and acids generally work best. Regardless of which electrolyte and which metals one chooses, a battery works because positive ions move in one direction through the electrolyte, effectively peeling off infinitesimal bits of the metal from one pole and sticking them to the other. Electrons simultaneously move in the opposite direction. Alessandro Volta, who made the first battery in 1799, used stacked copper and zinc plates as his metals, each separated by paper soaked in brine (the electrolyte). A nickel-cadmium battery uses nickel oxide hydroxide and metallic cadmium as its terminals (metals), and potassium hydroxide serves as its electrolyte. And one kind of lithium ion battery—there are many chemistries in play for this type of battery— uses lithium metal and manganese dioxide separated by lithium salts. What is important is that the whole process is a circle: as ions speed

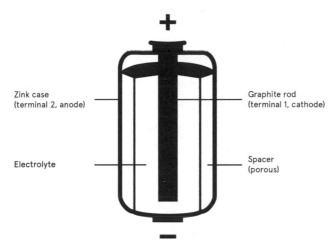

FIG 6 A familiar-looking battery, in which one metal (here, zinc) is used to construct the case and the other (here, graphite) runs through the core of the device. These are separated by an electrolyte and a porous spacer through which positively charged ions can travel with ease. To draw power from a battery it must be wired into a circuit—its positive and negative poles connected by a wire—like those shown on page 33. (*Loïc Untereiner*)

across the electrolyte, electrons travel across a conductor. Without the wire giving the electrons someplace to go, the ions don't move much, and vice versa.

This is why a battery can be made, sit in a package for years (though not indefinitely), and still be useful when popped into your TV remote or hearing aid or electric car. Though it won't work forever; the electrolyte solution will get tired, and as one of the metals is slowly eaten away while the other grows heavier, the battery lags and then ceases to work at all. A rechargeable battery reverses the direction of flow, using electricity from an outside source to move most (but not all) of the ions and accretions back to the other side, where they will be ready to flow back again when conditions demand.

The components of a battery, both metallic and electrolytic, are quite flexible. Different combinations of elements, however, give different results. Some produce more, or more constant, power, some last longer, some are lighter, others recharge more effectively. Every battery on the market is a compromise between price, toxicity, reactivity, and weight. There are some very effective, very poisonous batteries, and some great not so poisonous batteries that are insanely expensive because one element is rare, and there are some cheap nontoxic batteries that don't produce much power. A "revolution" in battery technology thus is less about inventing a new way for a battery to work than about tweaking any number of relationships between metals and electrolytes.

Lithium-ion batteries offer a particularly appealing mix of accessible and affordable materials with longevity, the fast, even release of stored power, and lightness. Initially, lithium-ion batteries were used mostly to power small computing devices like smart phones (in the early days, as one might recall, they had a tendency to cause laptops to burst into flame), but as time passed and we got better at making them, they got smaller, lighter, safer, and longer-lasting. Like their brethren, lithium-ion batteries can be bundled, and thus a lot of them can be used to store more electricity than just a few. It is now possible to drive a sports car or even a fancy sedan that runs entirely on lithium-ion batteries.

Newly on the market, and not yet in wide deployment, though it has generated significant buzz, is a home-sized lithium-ion battery storage system made by Tesla Motors that follows the same basic principles as the battery they put under the hoods of their cars. Called the Powerwall, it is both a practical and bizarrely lovely device. As if someone who specialized in 1950s refrigerator design was asked to picture the future of electricity, circa 2015. Very shiny and offered in an array of colors, the Powerwall, like Tesla's cars, was meant to be a crowd-pleaser. No longer will off-the-grid types need a basement lined with old-style, acid-leaky automotive batteries in order to watch TV in the evenings.

The promise of the Powerwall lies in its ability to change things for those of us on the grid, in two distinct ways. First, an affordable, long-lasting, easy-to-use home-battery system might enable solar power producers to keep their excess daytime electricity for nighttime use, rather than asking the grid, and the utility, to deal with both their over- and underproduction. Second, the utilities are exhibiting true hints of interest for using a mass of deployed, distributed battery systems as grid-scale storage. Vermont's Green Mountain Power will be offering Powerwalls to customers at a very reasonable price (about forty dollars a month) if the customer agrees to share the battery's storage capacities with the utility. It will serve as host to both home-made power and grid-made power, for rainy days and long, dark nights.

Many people who care about grid reform don't see electric cars as cars so much as great big batteries on wheels. Customers, when contemplating the purchase of an electric vehicle, tend to care about its capacity to move them speedily and reliably from one place to another. The grid guys, on the other hand, see an ingenious form of storage that doesn't rely upon the quirks of geology or climate, that blends seamlessly into its environment, and that can be made to work for tiny grid imbalances as well as big ones. For pumped hydro storage you need hills, for molten salt storage you need some serious sun, and for compressed-air storage you need salt caverns or similarly carved-out features. But a battery doesn't need to be any place in particular to work. The mobility of batteries, together with their scalability and relative effectiveness, has always been their charm.

With electric car batteries, if there were enough of them and if they were designed to both give to and take from the power grid (a capacity called V2G, for "vehicle-to-grid-enabled") peak load could be smoothed out to a gentle rise at the workday's end and variable generation might be balanced easily and thoroughly. Balancing solar in particular, which tends to export jitter to the grid every time a cloud passes overhead, tends to be an energy neutral process. Even in areas with a high penetration of rooftop solar over the course of a day the grid needs about as much energy in as it is capable of providing (out). As a result, a car's charge will stay about the same. With car batteries backing up the grid, we could have more green power, fewer polluting backup power plants, and no robocalls asking us to switch off the AC on the summer's hottest days. Rooftop solar could make all the power for desert and tropical metropolises without sinking the utility, overwhelming the grid, or causing the dreaded "solar duck conundrum" (more on which in a moment). These cars with their often-idle battery packs can be engineered to support and stabilize the grid by making their electric charge available whenever they are parked and plugged in. Even better, in only ten to fifteen years, expert predicters now forecast, most of us will be riding around in cars that drive themselves. These, too, it seems, will be electric.

Big "peak demand" shortages are not, however, energy neutral processes. In order to absorb the bump in demand that comes at the workday's end, without firing up any new power plants, all the cars—and there would need to be a lot of them—would be drained a lot. Not all the way, obviously, because then you wouldn't be able to drive home. And, with enough cars, no one battery would bear anything like the brunt of the drain. This solution to peak demand sounds a little like Marxism. Each car, never exploited, gives to the grid according to its ability while remaining available to the grid to take from according to its need. Together, all our cars keep our common electrical system strong. And the grid, with their help, can at long last balance itself. Upgrades, maintenance, and necessary investments to keep it smart will still need to be done, but the cars, as integrated, dispersed, deployable storage, would go a long way toward increasing both the reliability and the efficiency of the infrastructure as a whole.

Like any good vision for a more perfect future this one is not without solid grounding in the facts of the matter. A normal car, Dr. Gorguinpour, Director of Transformational Innovation to the Assistant Secretary of the Air Force for Acquisitions, explains, is a "very poorly used asset," functioning for maybe 3 to 5 percent of its useful life. If you have a vehicle that is plugged into the grid and provides energy back to the grid whenever it's not in use, "now suddenly your asset utilization is 95 to 97 percent. It's basically not being used as either an energy resource or a mobility resource when you are getting it fixed."

Better still, cars are used the least at night. The mass adoption of electric vehicles would thus help to provide a significant nighttime load that could be discharged during daytime hours when these same vehicles are sitting in parking lots. In addition to helping establish nighttime load, peak shaving, and helping control jitter—those tiny second-to-second shifts in quantity and quality of electricity on the grid—the widespread deployment of car batteries would allow us to integrate as many renewables as we wanted: the cars would hold the excess energy these systems make until it was needed, at which point the grid would automatically suck that power back out.

This would be a boon most especially in places with a significant penetration of rooftop solar and where there also happen to be a lot of cars. Every solar-powered city in the desert suffers from a particular problem—the inevitability of dusk. From the moment the sun rises, its rays shine down hot, powerful, and as productive as can be. The solar panels that, from above, in some places seem to have painted the city in glass make efficient, clean electricity by the megawatt. The city runs. All is well. Until dusk.

This moment has always been difficult for the utilities because it's when "everything is on." Some people are still at work and lots of people are already home. Now add to this jump in demand a radical fall in supply. The sun is going down and with it a precipitous fall in generation just as demand is rising with equal ferocity. This evening peak is much harder for a utility to deal with when the sun is powering more than about 25 percent of the grid.

"For an illustration look at Hawaii," writes the *Economist*. "On a typical sunny day the panels on consumers' rooftops produce so much electricity that the grid does not need to buy any power from the oil fired generators that have long supplied the American State. But in the morning and the evening those same consumers turn to the grid for extra electricity. The result is a demand profile that looks like a duck's back, rising at the tail and neck and dipping in the middle."

It is no longer the owl of Minerva that flies at dusk, but the solar duck which rears its ugly head as the sun fades away into darkness.

To solve this new kind of curve, the utility doesn't need twenty-four or even twelve hours of storage capacity; they need about six hours' worth to get them from four P.M. to ten P.M., when people start slowly trickling off to bed. Southern California Edison is hedging even this six-hour storage window a bit with its four hours of battery-in-an-office-tower storage. Most new solar trough plants also often include a modest level of molten salt storage, about six hours' worth, precisely because this is all you need to lop the tail off the duck.

The cars, however, would equally calm this curve. Forty-one percent of the electricity we use in the United States is used by buildings. If an electric car, which is essentially a big battery on wheels, were to make itself available to whatever building it happens to be adjacent to, be it a home or an office building, the owners of this structure would not need to invest in their own storage, nor would the owners of the grid. Demand shifts with people, so that as individuals leave work and drive home they take their electricity box with them. They are now using no power in the office (because they aren't there anymore) but lots of power at the house. That's fine, because the car is plugged in exactly where it is needed. As the battery runs dry in the late evening, household demand is also dropping as everyone is turning off their machines before trundling off to bed. Simultaneously electricity prices are dropping at which point the car begins to charge itself back up for the morning commute.

Notice that though electricity in this scenario is still public, storage has been thoroughly privatized. The utilities don't own the storage. Because you bought the car, you do. Electric cars thus also help solve the problem of who will pay for stabilizing our grid. People who own cars will. In the United States the proposed means for dealing with this oddity of privately owned, modular infrastructure is the "money for nothing" scheme. Utilities will credit you time-of-day rates for the wattage they suck from your car during the day and you will pay, theoretically lower, time-of-day rates when you recharge it, usually at night or during moments of over generation of renewables at whatever time of day. In this way, owning an electric car becomes like owning a little money factory. All you have to do is make sure it's always plugged in and the algorithms do the rest. At the end of the month you get a check in the mail. This admittedly has a certain appeal. It's also largely the same system that led to the rampant adoption of rooftop solar. You make electricity and either pay nothing or get a tiny check every so often from the utility that is buying it from you. It's a good investment. And, up till now, it doesn't work.

Denmark, which currently has 53.4 percent renewable energy on its grid, with a goal of 100 percent wind power by 2050, is also looking hard for a storage solution. Initially they bet heavily on the cars. The incentives were different, though equally well matched to the cultural and economic particularities of the place. First, if you bought an electric car you paid no taxes on the purchase. Normally, the Danes tax a new car at 100 percent, so buying electric was essentially the same as getting your new car at half price. But people still balked at the prospect. They complained, wisely, that if the battery is being drained and filled all day, every day, it isn't likely to last very long. In response the state instituted a battery swap system, so that whenever a car battery ceased performing up to spec, any Danish electric car owner could have their old battery pulled out and a new one dropped in for free at a filling station. In this way the battery is good for the life of the car.

Like the American scheme, this was also a good idea and it also didn't work, in part because neither the cars nor the batteries were good enough yet. According to the Danish climate minister (yes, they have one of these), the cars themselves are not ready: "We need longer range and lower prices before this becomes a good option," he said. "Technology needs to save us." Even at half the price, the Danes didn't want to buy a crappy car with a short range and a long charge time—imagine, you stop on the interstate to refill your car battery and it takes six hours. And, given how tiny Denmark is, what assurance is there that when their car battery dies in neighboring Germany or Sweden, gas stations there will be obliged to perform the required swap as quickly and at the same (zero) cost? Technology may be one problem, but so are borders beyond the bounds of which different laws apply and different incentives hold sway. This is as true of California as it is of Denmark. People don't want a car that forces them to stick close to home, no matter how good that car might be for the environment, the grid, or their pocketbook.

Add to this that if the vehicle-to-grid system is going to work as promised, all electric cars would have to be plugged in whenever parked. This makes for a massive investment in infrastructural rebuilding. How do you get a charging station into every spot of a cement parking structure without tearing the thing down and rebuilding it? What about on-street parking? Or private garages? If it's hard to persuade citizens to bear the brunt of the cost of buying these cars for the good of the grid, it's even more impossible to imagine convincing a city like Los Angeles to pay to wire itself up such that the grid is remade into a capillary-like system that permeates every place that any of their 8 million cars might come to rest. Wireless charging pads (or a similar device built into a nearby wall) might bring down the cost and increase the feasibility, but this technology too, while extant, is not yet ready for the mass market. Nor is it interoperable with the cars.

The expense and disruption of such an undertaking, given the current state of technology, would be epic. Even where most earnestly imagined, vehicle-to-grid storage systems aren't being realized. In part because we aren't buying the cars. (Norway being the exception, with almost 30 percent of 2015 sales being electric; even they are not yet deploying their wealth of batteries to the grid's benefit.) In America to date, only 1 percent of the cars are plug-in electric vehicles (a whopping 3 percent in California). Of these, almost none are vehicle-to-grid-enabled. And the numbers shrink where they should rise: only about 0.5 percent of fleet vehicles—such as those run by the postal service, UPS, or any municipal government—are electric. The easiest way to begin operationalizing "V2G" technologies is to do it with fleets, all of which park together in the evenings, lessoning infrastructure costs for the installation of two-directional smart charging and also vastly simplifying the economics of figuring out how to pay a utility customer residing in one service area for the power she supplies to the grid, or takes from it, while parked in a different utility's service area. Fleets should be the first adopters and yet, they aren't.

If we are talking about the big grid, electric cars for storage remain every bit as much a dream as room temperature fusion for power production, or the air as a conduit for the long-distance, wireless electricity transmission. The world would be a better place if all these things did work. But, for the moment they don't, at least not at the scale, or according to the terms we want them to.

Nevertheless, every concrete plan for the adoption of variable generation at a rate higher than 30 percent is premised in part upon the "fact" that we the people will be buying electric cars by the millions "in 30 years." Perhaps we will be and perhaps we'll be charging them with fusion, but for now we aren't and we don't. This holy grail retains its mass only in the minds of those struggling to build the better future in which we hope one day to reside.

As appealing and impossible as this vehicle-to-grid scenario seems for the big grid, it is turning out to be a lot more feasible for smaller ones. Not the greater Los Angeles Light and Power District, but the microgrid that is the Los Angeles Air Force Base; not Washington, D.C., but the microgrid that is Joint Base Andrews in nearby Maryland; not all of New Jersey, but the microgrid that is Joint Base McGuire-Dix-Lakehurst, just outside Hanover. The next time something like Superstorm Sandy hits the East Coast, Hanover will lose power, just as it always has, but the military is banking on the base, the town's nearest neighbor, remaining fully, resiliently operational through a mix of microgrid technologies including solar panels, fuel cells, diesel generators, backup battery systems, and a small fleet of electric cars.

The military is in a unique position, for as it converts its U.S. bases and foreign forward command units into an archipelago of microgrids it comes to be not only the sole owner of all the electrical infrastructure in its domains, but also the sole owner of almost all the vehicles. In the field, converting all nontactical vehicles to battery power further reduces the supply chains for liquid fuel; at home this same transition makes their microgrids even more robust—in everyday use and in times of crisis. As a result, the DoD, which operates a fleet of 200,000 nontactical vehicles, is working to convert them all to electricity with vehicle-to-grid technologies designed in from the start.

To date, only the Los Angeles Air Force Base has turned these ideas into workable technologies—though the same transition is planned for the other bases mentioned. So far in L.A. they have forty-two vehicles, some cars, some trucks, and a big van. Six of these are hybrids, but the other thirty-six "have the capability to direct power both to and from the electrical grid when they're not being driven." As such, it is, according to the air force's Dr. Gorguinpour, "the largest operational V2G demonstration in the world." Even these few vehicles can, in a pinch, provide more than 700 kilowatts of power to the grid (about 150 houses' worth). Less dramatic is the underlying motivation for the project: "The vehicles enhance the power grid's reliability and security by balancing demand against supply without having to use reserves or standby generators."

The power grid here is both the air force's power grid and our own. When islanded in times of crisis, it's just their microgrid that will benefit directly from the balancing capacities of an electric vehicle fleet and the "unique" charging stations that allow these to feed into and draw down from the grid. Most of the time, however, their grid is indistinguishable from our own; technically speaking they are one in the same. The electricity they generate does not stay on the base; rather, it flows into the common power supply. And the balancing done by those thirty-six vehicles the 95 percent of the time that nobody is driving them is being done for the grid as a whole: Los Angeles benefits from the thirty-six, California benefits, the desert states of the Southwest benefit, and the sodden states of the Pacific Northwest benefit. The whole of the Western Interconnection is one system with many potential islands—more with each day that passes— and most days, none are islanded. The underlying texture of the macrogrid is changing, but in practice this has only minimal effects on its everyday operations. Most of the effects it does have, however, increase the reliability of this larger system rather than decreasing it.

Batteries, especially V2G-enabled car batteries, may not turn out to be the answer, despite the constancy of their hold over the minds of today's dreamers. That we are looking hard for a grail, sketching out and even prototyping a lot of different ideas, is itself an acknowledgment that we stand at the edge of something. Not the end of an infrastructure, though it could go that far, but certainly at the beginning of a new century's imagination of how that infrastructure might be adapted to suit us better.

Regardless of how one looks back at the twentieth century there is no way to imagine it without the gradual and complete adoption of electricity. There are other infrastructures, equally important, but as we slip from the industrial age into the age of information electricity's place at the heart of our life and culture only increases. It is money, it is data, it is computing, it is algorithms so complex they begin to border on intelligence. Electricity powers this world, but it is also the means by which information constitutes it. The infrastructure we need to hold our present in place and allow for its increase is not the same grid that provided the twentieth century's national uplift and union.

It is right, then, that we now live in the era of infrastructural dreaming and that these dreams would be centered around finding ways to unplug electricity from the system of wires and power plants leftover to us from the past. Storage, whatever forms it will take in the end, is not the holy grail because it helps to balance the grid we have (though this is the story we like to tell ourselves). It's the holy grail because it allows us to build an electric world that functions otherwise, that has the flexibility to move and change with whatever the twenty-first century will throw at us. Or, more correctly, whatever remarkable, impossible things we will build into our own near future.

It may sound odd to advocate for giant hydroelectric storage pistons on German plains, but why not dream big? And dream diversely? Dreams, even ones sketched and prototyped, are the way of the future, the grail is its driver, and today's entrepreneurs, and kooks, are not so unlike the inventors who, at the end of the 1800s, reached out and invited that future into the nation's living rooms and offices, with some successes and many failures.

Batteries have their place in all of this, but it would an error to imagine that their centrality in our imaginations over the last five years will necessarily place them at the center of our twenty-first century grid. They might be. But then again, it—the grid itself—might not be. In all of this, it is only important to keep one's eye on the grail. What are its parameters? What do we dream of? What do we seek? And how will this thing, should we invent our way to it, change everything about the circumstances that gave rise to the dream to begin with? "The challenge right now," architect John Keates enjoins us to consider, is that given that we don't know the answer to what comes next, to also ask: "what are the ethics that we set for ourselves? And to be aware that when we venture out into untrodden territory, that we are able to ask that question and dare to act when there is no clear answer." The dreamers and builders and kickstarters of electrical storage campaigns and components are doing an admirable job of this.

CHAPTER 9

American Zeitgeist

In German the word for an addict is not like our own. In English to be an addict is to have a totalizing identity, controlled by a "soul-destroying, mind-numbing obsession that makes normal functioning impossible." As such, addicts are wholly governed by the fact of addiction, which, to put it mildly, "obscures other important qualities about them."

Not so in German. The most accurate, if graceless, translation for an "addict" would be something like "seeky." So that if Americans are addicted to oil, then our own language renders us nothing but. By this definition, all we do, all the time, is suck up oil in ever vaster quantities, at ever-increasing cost, and at the expense of our general well-being. Many people in the United States, as well as citizens of Kuwait, Iraq, and Venezuela, would likely say that this is an apt description of Americans' behavior in relationship to their oil. Using German logic, however, we might propose instead that Americans are "foreign-oil-seeky," just as heroin addicts are "heroin-seeky" and those poor souls addicted to *Friends* reruns are "*Friends*-seeky." This allows one to be both fairly obsessed with one thing while simultaneously interested in, and capable of, many others. What is curious about this Germanic way of typifying addiction is how well it captures our everyday relationships to electricity. We are deeply and abidingly electricity-seeky.

You've seen it yourself. A woman enters a coffee shop and pauses. She does something that looks rather like sniffing the air, except she is using her eyes, sometimes there is slow stroll involved, until somewhat unexpectedly she plops down into a chair, usually not the best one, and begins removing the contents of her bag. A phone comes out, a little pad of photocopies, a computer, and . . . a cord. She bends over, nudges an old man with a big coat out of the way, snakes her cord over his head and around the corner of a post, and then plugs it in. She was looking for an outlet. She is not an addict, not exactly, but most certainly she is electricity-seeky.

One sees the same behavior when a well-pressed businessman is found sitting on the sullied carpet of the airport boarding area waiting for his flight to be called while his phone charges from an ill-placed baseboard outlet. Or when, upon entering a cubicle for a meeting with its inhabitant, one finds him beneath his desk, khakied butt in the air, snuffling around like a pig for truffles, trying to figure out what's gone wrong with his surge protector.

Electricity moves us, if not yet in big ways and over long distances: cars can hardly be said to run on it, and jet planes simply do not. On a smaller scale, however, it is constantly orienting us, inciting in us a daily seeking behavior; we subtly structure our lives around access to outlets, though we see evidence of this fact only when the outlets have been poorly placed. Design failures as much as outright blackouts are what transform our everyday "seeky" into active and obvious seeking.

Even the most professional among us lose decorum as we are about to lose power. The richest maintain members of their staff, and circulate though specially designed architectures, to ensure that such inconveniences never happen to them. They, too, are electricity-seeky, they just outsource responsibility for the addiction to the world around them, making it seem as if the phone of privilege always holds a full charge. This magical relationship of invisible ease, of forgetfulness, of electricity without "seeking," is in fact what we all long for from our grid. Infrastructure should, according to the design guru Donald Norman, fade into invisibility. It should be made to disappear from sight and equally to disappear from consciousness. It should be quiet, task-specific, and unobtrusive. We shouldn't notice it, we shouldn't think about it, and we shouldn't seek it.

In all of the talk about what will make our current electric grid into a better system, usability rarely enters the discussion. It is easy to forget that there are two ways to look at the grid, from above and from below. What we see when trying to grasp it more globally are the relationships between technology and finance, or between legislation and cultural values; we see the grid for its vastness and complexity. We try to grapple with it as a system the expanse of which, geographically as well as historically, is truly outside the capacity of the human mind. Gaining some measure of purchase on this enormity has been my task here too; I don't think we can "get" the grid without an intense care for the cultural context and microdramas that are playing out between technologies, people, government policy, and corporate concern for the bottom line all the time.

Be that as it may, this is not how most of us interact with the grid. Most of us are not the Straws, or the U.S. military, or New Jersey Transit. Most of us still don't have solar panels or an intimate relationship with batteries any larger than those that power our phones. Our grid, the part of it we know and interact with, looks like the cords, plugs, outlets, and switches that link our portable electronics to the wall. This, too, might be thought of as the grid edge, since the moment that we unplug electric power from the big grid and carry it around in our pockets is literally where the grid disaggregates, becoming something different and new. The charger is the final cord, the one we know best, and the smallest dendrite in a world of complexity.

In this, the last chapter, I'd like to turn toward our more intimate and personal interactions with the grid and our desires for these. These interactions, and wants, have a pattern, and this pattern, I suspect, exerts more force over the shape of grid reform than experts, or those with a more global view, give credence to. What we would like ultimately is that our infrastructure move us less and have a less obvious presence in our world—not just visually, not just when it breaks down and emerges forcefully into consciousness, but all the time, and in both little and big ways. No more outlets, cords, or plugs would be nice. No more blackouts, short or long, would be great. So would a return to an earth with a stable climate, a planet that grows neither warmer nor colder because of our attachment to fossil fuels for making power.

In short, we'd like our grid to whisper away, to be less devastating in its effects, and to work without deputizing us to the process. We'll keep electricity, thank you. In fact, the further we proceed into the age of information the more electricity becomes the base for all that we do, from banking, to reading, to collaborative thinking. The future promises an even more thorough integration of electricity into our lives, more data (which is after all, just electricity), more "smart" things (coming to populate the Internet of Things), and the elimination of fuel from cars, necessary if we'd like to stop global warming before it exceeds the 2-degrees-Celsius disaster line. Most important, we'd like this means of "being electric" to come from nothing, to be transmitted by nothing, to cause no damage, and to work always and wherever.

This abiding cultural attachment to electricity only makes the unwieldy ways in which we have to move in order to access it all the more salient. If only we could dematerialize the infrastructure while simultaneously making power ambient—ever present, never sought—then perhaps we'd have an electricity system better suited to the present and better oriented toward a future that meshes with the data-driven and data-dependent beings we are becoming.

The question is, of course, how to do this. Especially since our everyday desires for the future of electricity have relatively little to do with what the vested interests pursue in their attempts to maintain the technological and fiscal viability of the current system. One of the problems with the grid is that we can't just shut it off for a couple of years while we come up with something newer and better. We need it to be running all the time. One recent author described the project of overhauling the grid as akin to "rebuilding our entire airplane fleet, along with our runways and air traffic control system while the planes are all up in the air, filled with passengers." Systemic reform under these conditions is not easily accomplished: anything we add to the grid must have the capacity to interface effectively with everything that was there before, while everything we subtract from it must not disrupt the flow of power that we are so reliant upon.

Given all this, the best routes forward are those that take the mess of competing interests seriously and design for them. This range of interests should not be limited to investors, visionaries, legislators, utilities, regulators, and all the other folks that have an active interest in the grid; it also needs to include people who don't care a whit for the business of electricity.

Figuring out how to design a system for maximum inclusivity is harder than it sounds, in part because it's difficult for any one player in the giant tangle of our grid to have much comprehension about what motivates the others. As the near meltdown of the Davis-Besse nuclear reactor or the incessant gnawing of squirrels makes abundantly evident, it's challenging enough for each of them to be masters of their own domain. Yet, when everyone has their eyes pressed tight up against the demands of their own jobs, it's the people, the payers of bills, who remain relatively easy to ignore. Unless we make trouble, with guns, air conditioners, or home solar systems, we remain a relatively lumpen mass lacking even basic demographic nuances one might expect any twenty-first century business to employ. For example, in all my research into the grid I have never heard a utility customer referred to in the feminine. When speaking of the users of electricity, the (mostly) men who make the system work, and the (mostly) men who push at it and try to invent a way beyond it, imagine a nation of users who are also men. This is necessarily only ever half true. But if the quiet but undeniable fact of gender has not percolated up into the consciousness of those who make, and remake, our grid, what else is being lost? Attentiveness to the details, and not just aggregate data, is critical to the effective reform of an infrastructure so essential to our lives. Yet asking everyone to practice a more nuanced awareness of ecosystems complexity is, I suspect, more than they can bear.

There are thus three problems of different kinds that meet at the grid and get stuck there: How to deal with the combined interests of many different players—which does, and should, include global warming. How to deal with the legacy technology, which is to say the grid we've got. And how to deal with the fact that it's made and run by humans, who are by their nature rather squirrelly and shortsighted. Rather than letting it all go wild, giving everyone a blunderbuss, a dedicated banker, and an agent of the press—the way the grid was won in the late 1880s—a more practicable solution would be to design something radically integrative. To err, at every moment, toward inclusivity and to design for the easy incorporation of as many different interests as possible. This will mean a clear set of obligatory standards that twist the arms of even the most stubborn players toward interoperability. It will probably also require legal and regula-tory intervention. Plus we will need to find a way to pay for the most basic elements of the infrastructure, the wires and poles, that few people care about and yet must stay standing and in good condition for all the rest of the grid to work at all. Investors' money may flow toward newfangled forms of generation these days, and toward private companies, whether inventing microscopic components or repurposing salt caverns, but basic maintenance is still on the back of the utility. We, the users of electricity, pay those bills, until we don't.

Coming up with a good system for grid-scale storage, with its capacity to unlink generation from consumption, is one way of pushing the grid toward a more open and assimilative attitude, but it's not the only way. A second, increasingly popular means for translating interests of different kinds into a single system is to rely upon a platform—an integrative computer program rather than a gadget. In order to help ensure that our grid is wrenched out of its current workings, this platform would need to be open to all the strange sorts of things people are dreaming up and building today (from vehicle to grid-enabled self-driving car pods to real live nanogrids) *and* to the boring old stuff we're stuck with for the moment (like natural gas combustion plants and old coal or nuclear), and also to the desires and activities of regular people. All without letting the basic structures of the grid get too rotten or out of date.

A platform is an interesting tool to think with in part because it moves us into a domain where computing, or "digital" systems, becomes the means for solving mechanical or "analog" problems. Platforms literally use computers to make messes operable. Uber is a platform. It takes drivers driving around and organizes them into a means for nondrivers to also move around, in the same cars. It brings existing resources together and creates a functionality where once there was only traffic. Facebook and Twitter are platforms for sharing information and opinions across vast social networks. In the process they produce actively intertwined relationships between strangers as well as friends. These ties, unexpectedly, move news more quickly than more traditional media and thus allow for concerted action and organization significant enough (at times) to bring down governments. This pattern of using platforms to organize unrelated and competing interests into new social and economic formations is a comfortable one now, in America. Comfortable enough that some already exist for our grid, and a solid subset of people inside the system are working out how to make these even better at integrating absolutely every-thing, and even some "nothings," than they already are.

Within this push for integrative grid reform there has arisen a curious driver—a certain affinity with zero. In weird corners and in main-stream parts of the country one finds people advocating for aggre-gated nothings, for no fuels, no wires, and no measurable effects. Conservation and efficiency, ascendant in the 1960s and '70s, were the precursors of a valuation of what wasn't used and not needed. And the debate that rages today over how one might accurately count a watt saved, or value and remunerate a power plant not built, also began back in the days of President Carter and his Cardigan Path. In the age of wireless communication systems, we can add another vision: How might we make a grid less material rather than more so? Across domains one begins to see an abiding concern for ways of reducing the material impacts of infrastructure and counting every zero we make as if it were something substantive. Even when not conscious of this trend, a good many people, companies, and interest groups are leaving to value what is not there as much as what is.

The greens, to take the obvious example, root sustainability in getting power from nothing and producing no waste in the process. Renewables, once they have been built and put into operation, are not chemically polluting and they involve neither extraction nor waste—nothing is brought up from below ground, nothing is burned, nothing is boiled, and nothing is released to thicken our atmosphere. Engineers, for their part, are in favor of almost any system that can increase the amount of power they can get from a fuel source. Doing away with heat engines, with their inevitable thermodynamic limita-tions, is for them a big step in the right direction. Large corporations with significant investments in the energy sector like the fact that the cost of fuel can be removed from the power-production spreadsheet of inevitable expenses. For every watt not made from coal or gas or plutonium these zeros are produced, and they add up. It's simply easier to make a profit if one can reduce to nothing the cost of a necessary ingredient.

These motivations for supporting green energy are believable and defendable in their own right, but they also all uniquely conform to the spirit of the times: renewables use nothing to make electric power just as phones use the Internet and 4G networks to produce information from thin air, just as wireless charging stations for home electronics are now being integrated into the furniture we normally put them on anyway. Indeed, consumers, who have gotten used to getting information, music, and even movies from thin air, and who will soon be maintaining an electric charge in their batteries through similar means, are increasingly reluctant to embrace anything that has the aesthetics of an older, more materially invasive system. And though consumers are not the arbiters of the grid, their predilections do affect it.

The grid's wires bear the brunt of this discontent with the materiality of infrastructure. Bruce Wollenberg, a power systems engineer at the University of Minnesota, explains the utilities' frustration in trying to add more high-voltage lines to the nation's transmission system, saying, "People don't want power lines—period. They don't like the way they look, they don't like a lot of things. It's universal across the country, and I think across the world. People don't want more power lines." Similarly, Caltech's Nate Lewis, one of many engineers working on artificial photosynthesis, speaks with the same reiterative stutter. His team's artificial leaf has "no wires. I mean what I say: no wires. Leaves have no wires. In come sunlight, water, and CO_2, and out come fuels." In both cases it almost seems as if the presence of wires or lines in a new product will sink it.

If consumers are unwilling to allow utilities to build wires across private as well as public lands—as Not-In-My-Backyardism rises to greet the grid—they are happy to pay more for a night table with a wireless power charger built in, like those Ikea is currently rolling out. And, somewhat surprisingly, it is not just about the scale of the thing one is paying for. The utilities have been quick to recognize the fact that people who are normally quite stingy with their electric company, including those actively opposed to new high-voltage wires, will voluntarily pay a surcharge on their bill for renewable power. This surcharge is usually offered as a "percentage of total consumption" with deals like 85 percent wind or 100 percent renewable (wind, solar, hydro).

When Xcel, the maligned former utility of Boulder, Colorado, offered this deal to its Minnesota customers, one friend of a friend on Facebook raved: "If you are a homeowner, there is no reason on earth not to do this . . . it will cost me $4/month and my energy from here on out will be 100 percent wind powered. What a great feeling! Plus, the energy that supplies Windsource will be purchased entirely from wind farms in Minnesota and will go above and beyond government mandates."

What matters here is the emotional force behind the idea for a certain kind of customer. No one connected to the grid, even if they pay the surcharge, is getting 100 percent of their electricity from "immaterial" fuels. Nor is their power, on a grid that spans half a continent, guaranteed to be more local. The same terms may be used to laud an environmentally friendly, locally produced electron as an organic, locally grown tomato, but the two entities are impossibly different. Both a customer who has checked the renewable fuels surcharge box and her neighbor get the same electricity. The principal difference is that the electric bill of the "Windsourse® for Residences" customer runs her four dollars more every month.

This paying more for locally sourced green electricity is not a utility scam. The surcharge does usually factor into a utility's budget for the immediate purchase and longer-term investment in renewable power. You may not get to use 100 percent wind power, but you do subsidize the increase of this means of making electricity for everyone on the grid over the long haul. The fact that this choice is available allows individuals to take a measure of fiscal responsibility onto their own shoulders as a means of committing to immaterial fuel sources that make sense to them.

The argument, then, that our grid wouldn't be in such a dire state if only people were willing to pay more for their electricity is largely moot. People are willing to pay more, but only for certain things. The less solid these things, the less visible, and the more thoroughly integrated into the built environment they are, the more likely individuals and companies are to volunteer their money for the cause. The ways in which these immaterial power sources also fit into the rising tide of concern for "green" energy makes the utilities' job easier. If renewables will help raise revenue, then renewables will show up on the bill.

As big wasteful things give way to their smaller less wasteful brethren—as coal plants have given way to wind turbines in Minnesota—it won't just be the fuel that's whispering away from our grid. The incandescent bulb, already illegal in much of the world because of its incomparable wastefulness, uses only 5 percent of the power it consumes to make light; the other 95 percent comes out as heat. This familiar and beloved bulb has been largely replaced, first by the compact fluorescent bulb, which pretty much everybody hates, and more recently by LEDs—small, intensely efficient, long-lasting diodes. Or take the refrigerator, the household's second-greediest device (after the air conditioner). It has been the object of intensive R&D since the late 1970s. Today's fridges use about a quarter of the power they did in 1975. This increased efficiency is true of all refrigerators, but for a wee bit more you can get one with an Energy Star rating—a truly good fridge, in both senses of the word. Just like today's refrigerators, our dishwashers, clothes washers and dryers, lighting systems, even newer generations of computers are designed to function equally well with much less of a draw. All of these machines look like and work just as well as their precursors. They just use a lot less electricity (and water) to do the same job.

And while some people, mostly those older than forty, still primarily approach problem solving in terms of buying better things, the nonbuyers of things, a characteristic the millennials have become famous for, take it one step further: Why buy a bulb at all? Why a fridge at all? What might we do to render the bulb and the fridge obsolete entirely? Might not a room be lit by a wall woven of fiber optic cabling controlled from the phone? Might not the fridge be reduced to a ceramic container and a shelf with a cold spot built in?

One can continue in this vein: Might not a solar panel be made as a window or as a roofing tile or a bit of roadway or a tree? Might not outlets be replaced with radiant electricity? The shower with a vapor stall that wets the user just as well with 30 percent of the water? And so on. These are not idle queries, nor should they be easily dismissed. This generation is already in the process of embracing technologies that render obsolete the light switch, the door lock, the car ignition (and its key), and the credit card. Why not the fridge, the outlet, the light socket, the showerhead, and the ineffable, uncontrollable, proliferation of chargers and their cords? The Clapper might still be the means for remote-controlling the lights with the most significant market penetration, but this is unlikely to be the case for long.

All of these changes—to appliances, light, and heating and cooling systems—are happening in the home, our most intimate space for thinking about and using power. We heat where we live and we cool it; we eat and cook, wash and dry, cruise the Internet, play video-games, watch TV, and listen to music there. On good days we even remember to charge our portable electronics before we leave for the day. In the aggregate, what we choose to do in and with our homes matters to the grid. This much we have learned from the home solar movement, but conservation and efficiency—ways of causing power not to be made—matter just as much, if not more, to the future well-being of our energy system.

If one steps back, it becomes clear that changes to the domestic sphere are but a small, if necessary, element of a systemic transformation. Offices, factories, and other workplaces also need to be retrofitted and outfitted, and so do the places where we shop, socialize, and eat out. These are slower to transform, largely because of the cost. Putting solar panels on the roofs of all the nation's Walmarts (this is happening) is not the same as putting solar panels on top of one's garage. Nor is the difference just a matter of scale.

Every suburb and semirural outpost where big-box stores cluster has its own culture of wires, its own utility, balancing authority, and regulatory apparatus that must prepare for (and often upgrade existing infrastructure to accept) every new source of power. When we consider the grid in this way it becomes even more evident that some sort of hub capable of translating across all the competing interests and integrating all the structural intransigencies will be essential to the success of an infrastructural upgrade that is national in scale.

In proposing better means for making and delivering electricity, we need to ask ourselves: Does this path, the one we labor to produce, the one we legislate, the business plan we follow, cut off a whole set of options, or does it allow these to wrangle on in there with the rest? Ease of governance needs to cede way to a means of organizing a diversity of interests around a single vision. This is difficult even when the vision is fundamentally about implementing systems diversity. In the high-stakes, low-sex-appeal battle to ramp up the interoperability of the grid's thousands upon thousands of subsystems, the most boring, if heated, conversations behind the scenes focus on standardization. A platform is not just a software problem. We actually need the technologies that currently constitute our grid to be able to work with, and communicate with, newer components and newer ways of doing things.

The list of standards necessary to make cross-generational interoperability even possible is dizzying. These include items like CEA-852.1:2009 "Enhanced tunneling device area network protocols over Internet protocol channels" or C37.118.1 "Standard for synchrophasor measurements for power systems" or IEEE 1588 "Standard for time management and clock synchronization across the Smart Grid for equipment needing consistent time management." Gone are the days when a simple armature could bring all the disparate devices functioning on our grid into a new resolution. Achieving a "universal system," the final form of which we can neither imagine nor plan for at present, is going to take a great deal of legwork of this kind.

While teams of career bureaucrats struggle to devise standards that will make a heavily networked grid functional, other folks in other places take steps to ensure that these won't ever be needed. The wise grid, as this option is known, has a thousand opponents, each arming itself in whatever ways it can. Effective systems change can be derailed by any of these. If everyone with a stake in the game chooses to limit, rather than work with, the chaos of the present, we will end up with a balkanized system built of roadblocks and blind alleys. Any action, no matter how small, against interoperability creates new hurdles for anyone hoping for a future grid grounded in flexibility and reliability via diversity.

Even forward-thinking California has proved boneheaded on this point. Late in 2015 the legislature in that state passed an extraordinary new renewable energy standard into law, which (among other things) obliges California to make 50 percent of its electricity from renewables by 2030. As remarkable as it sounds, at the core of this piece of legislation lurks an unexpectedly retrograde logic. The only renewable electricity that will count toward the 50 percent is that produced by central stations. Rooftop solar will not be counted. This despite the fact that growth in homemade solar power has been exponential over the past half decade: 45 percent of the nation's total residential solar power is now in California, 82 percent of which has been installed since 2010. At present, rooftop solar is producing three times more electricity in California than are central station solar plants, despite seven of these massive power generators having come online in that state in the past three years. As this legislated reluctance makes clear, the utilities, which have a strong lobby in that state, would prefer to ignore small, distributed power producers; homeowners are too complicated for them to control (or even to bargain with), and homemade electricity is almost impossible for them to profit from. This particular set of players was also hoping that home installations would plummet after 2016, when the renewable energy tax credit was scheduled to expire, since this incentive has refunded 30 percent of the cost of installing rooftop systems since it was put in place in 2005. It didn't expire, but that matters not in California, since the legislation was already in place.

A conservative eye toward the future might see in this a landscape of power production slipping backward. If homeowners lose the direct financial incentive to contribute to public power, then perhaps central stations might remain "central" to the business of making money from supplying electricity. Turning an institutional blind eye to dispersed renewables—by refusing to let their power count—in favor of big wind farms and sunpower factories is a first step in a process of marginalizing the means of making electricity that many regular people obviously prefer. In this way the new California law has given utilities free reign *not* to work out how all the various means for generating, saving, and storing electricity might come profitably together. They have effectively limited systems diversity in favor of securing, more tightly, their own interests. The central stations may have changed their fuel source, but the California legislature has opted to leave them and the twentieth-century logics they embody at the heart of that state's grid.

The explicit goal of the legislation, of course, is to motivate the utilities to invest more heavily in large-scale renewable power installations. A sentiment difficult to fault, and yet, in failing to require the utilities to take full advantage of all the renewable power resources available to them, California's legislature has virtually assured that grid reform in that state will fall far short of its potential. That this blindness is directed at a technology with an evident appeal to the people makes it downright nefarious. Here, then, we can see how a good decision, in almost all terms, is also a bad one when considering the future-possible for our grid—largely because it excludes the desires of regular folks who are choosing to take their business to smaller, more innovative companies. The utilities are masters of ignoring what people want. And they are practiced in running competitors out of business by controlling the market. California's lawmakers, in this case, have given them a free hand to continue to do both. The right path was the more difficult one—to ask the utilities to work out a system whereby all renewable power was counted and integrated in the 2030 goal.

The problem of keeping the grid's wires alive, at least those we can't get rid of, will never be solved by sweeping the persistent problems under the rug. If Tesla Motors or some other smart start-up can create a battery pack that is sufficiently cheap and sufficiently reliable, people with the means to make their own electricity will start dropping off the grid. This might seem to solve a short-term problem for the utility, but it would significantly undermine a larger, national project of providing the same quality of electric power, at a fair price, to all America's people. For this we will need a grid capable of bringing together different value systems as much as different mechanical systems into a functional whole. Among the many things we must figure out in order to help make this happen is a way to allow utilities to make money off home solar systems, but even this possibility was curtailed by the California legislature's groundbreaking renewable energy bill.

Way back in the 1970s, Amory and Hunter Lovins's argument for the integration of renewables was grounded precisely on this means of effecting an infrastructural transformation worth its while. Yes, renewables are green, and on a polluted and warming planet that is clearly a good thing. More important to the Lovinses, however, was that these technologies, when connected to our existing grid, force it to work differently. It would become a more secure system because grappling with distributed and variable generation is such a mechanical and fiscal stress to the central station model. This potential for a better, more robust, more secure grid grounded in technologies we both use and like is made possible by letting people find ways to make and store electricity at the smallest scale without excluding them, structurally *or* legally, from our common system.

The grid, as should be clear by now, is not a technological system. It is also a legal one, a business one, a political one, a cultural one, and a weather-driven one, and the ebbs and flows in each domain affect the very possibility of success of any plan for its improvement. If the integration of systems across domains, especially the irritating bits, cannot be made to flourish, the problem will be not with the machinery we use or the technology we govern, but with us.

At issue is that as poor as the utilities are at accepting small reforms by small players, Americans, in general, are not especially practiced at ascribing a value to what is not-used, especially when the count is of something as abstract as a watt. And yet conservation and efficiency measures that reduce our need for electricity are as important to reforming our energy system as is the mainstreaming of renewable ways of making that electricity—large and small. Two different sorts of things, then, need to be integrated into our accounting. First, all the electricity made, no matter who is making it. And second, all the electricity not used, no matter who is saving it. If we can work out how to do this, systemically, it will start to matter when a couple of big-box stores, a cement factory, or a subdivision or two are energy-efficient enough that a utility, or anyone else, can avoid building a new power plant. The owners of these enterprises, just like any homeowner who's invested in a smart thermostat or a host of compact fluorescent bulbs, naturally want credit for what they are saving. Current rate structures, which charge different amounts for watts depending on when in a billing cycle (or when in a day) they were used, however, don't translate efforts at conserving power very well into the charges on the bill. The problem becomes, given that all these various customers are still on the grid, how the electricity they don't use—their so-called "negawatts"—might be counted, paid for, and deployed.

This is the real story behind contemporary grid reform: not just valuing electrons made by unusual producers, but valuing electrons that we never needed to make at all—the saved power that we shouldn't even notice has gone missing.

In 2011 I joined a green energy tour group of engineers and industry insiders traveling around San Diego looking at the various projects under way there. San Diego Gas & Electric is a notoriously forward-thinking utility. One result is that the city has some pretty interesting grid-related things going on. We saw some of these. An office complex with a solar panel on every available bit of roof, including the parking lot; a training facility for journeymen electricians with two empty car-charging stations conspicuously placed at the front. And, much to everyone's surprise, a suburban Albertsons grocery store powered by a boxcar-sized fuel cell out back by the Dumpsters. It's unusual to see a fuel cell, and the box drew plenty of *oohs* and *aahs* from our group.

From the store manager's point of view, however, it clearly paled in comparison to the energy-efficient, understated refrigeration they'd had installed. He tried, unsuccessfully if repeatedly, to draw our attention to his coolers and freezers, his windows and electric fans, his recycling system for cardboard boxes. These designs, machines, and systems all worked to lower the absolute amount of electricity needed by the facility. They were what made the fuel cell out back a viable technology for power production in the first place. It was not an Albertsons off-the-grid, but its draw from public power systems was close to zero most of the time.

An installation like this, at long last, brings the point home: if the experts and insiders on the tour were excited by how the watts that ran this Albertsons were being made, the store's manager was excited by all the ways in which these watts were no longer needed. What he saw when he looked at his grocery store was not a power plant but a machine for making negawatts; it was that machine that all of the rest of us had trouble seeing precisely because it looked, and worked, exactly like a grocery store. The lighting was perhaps marginally more pleasant, but otherwise, what had been accomplished was praise-worthy less in relationship to how the power for this store was generated and more in terms of the various ways that the whole thing had been redesigned not to need it. Or, as Amory Lovins (who coined the term "negawatt" way back in 1990) said, "Customers don't want kilowatt-hours; they want services such as hot showers, cold beer, lit rooms," and this can "come more cheaply from using less electricity more efficiently."

This way of thinking—of lowering consumption while minimizing the need for grid-provided electricity, does not simply conform to the spirit of the times. It helps to further it. The power of nothing is expanded, by undertakings like these, to include lighting from the sun, cooling from fan-made breezes rather than chemical air-conditioning, and wattage not used because it isn't needed.

These saved-watts or negawatts are the electric power a machine or a building or a lighting system or a factory doesn't use. Though a negawatt is a theoretical, rather than a real, unit of non-power, it serves the purpose of allowing us to measure and quantify avoided consumption.

Given this, the store manager's point is well taken: Why not do the same with less? In this we can see the remnants of the Cardigan Path. It is there in part. We can use less. In fact, we do use less. Electricity consumption in the United States ceased its increase in 2007 and is predicted to remain flat until 2040, despite infinitely more electronics and significantly more people. As important, however, are the changing presumptions that surround a life of modest energy consumption.

We needn't suffer privation as a result of using less. This is the message that Albertson's Corporation has built into this single store. It is equally the message of the shift to LEDs, home solar panels, and ideally to electric cars. In the twenty-first century, the Cardigan Path has been remade more simply into the Path. These days, not even Jimmy Carter needs to wear a sweater.

The Path means less energy consumption without turning down the thermostat, wearing a fleece, or lighting a fire. It means long hot showers, cold beers, and well-lit rooms. There will be no privation in the new world of less. (Not "less-is-more," mind you, but "the same with less.") At least, this is the goal of those committed to an accounting of negawatts and equally of the man who manages the most energy-efficient Albertsons in the United States.

Even ten years ago, the answer to this desire for "the same with less" was that saving kilowatt-hours was expensive precisely because retrofitting inefficient buildings with more efficient technologies is the *least* cost-effective way to achieve the goal. Negawatts simply cost too much to be worth their while. This is the reason why there is only one Albertsons with a fuel cell and a remarkable cooling and lighting system and there are literally thousands of Albertsons that rely on grid-provided power to run wasteful coolers and poorly designed lighting and HVAC systems. Retrofitting is expensive in ways that don't quite seem to pay off.

A case in point. I live in a very cold place, very cold, and one corner of our house was never insulated; it happens to be the corner that our bathtub drain runs through and so every year the drain freezes and we have no tub for six to eight weeks. Our solution to this would make Amory Lovins blanch. We just run the water in the tub all winter to keep the drain from freezing. We don't pay for water, it's not metered, so like much of the rest of our winter city of 2 million the taps trickle for months on end. Despite the evident ridiculousness of this fix, we won't be taking out a home loan in the near future to pay thousands of dollars to have our bathroom ripped up in order to install $150 worth of insulation into a single wall. We don't want to invest the necessary money and, even more, we don't want to deal with the mess and bother.

This is precisely why, as long as investment in retrofitting remains the main means of bringing about a negawatt revolution (or, in our case, a negagallon one), it won't happen. The efficiency and conservation measures necessary to insure a drastic community-wide drop in power consumption either need to be built in from the start or they need to be "plug-and-play," like Energy Star appliances.

This is half the secret, then: to design and to build places, things, and machines in ways that effortlessly and invisibly—from the end user's point of view—reduce consumption. In some cases this might best be accomplished by reconceiving entirely the thing being built: to take the fridge out of refrigeration, the air conditioner out of air-conditioning, the light out of the lightbulb. In other cases, it might mean producing an identical thing, like a laptop that runs equally well on a quarter of the electricity as the same model three years earlier or a building that so seamlessly integrates power-saving systems that even its most constant users would be shocked to know that they are moving through a massive negawatt machine.

The other half of the secret is to get the grid to do most of the work in this direction for us. We need to enforce a system that takes the messes of the present and orients them not just around a different kind of generation or a different kind of distribution but also around countable nothings. Efficiency, not just for the sake of it, but as a structured means of generating less power to begin with. We don't use what isn't made, and (this is the new bit) this non-use will get factored into our financial thinking about the grid, and its reform. If it can be given a stable price, a negawatt will matter as much to how actors big and small choose to reform our grid as a tax cut, a subsidy, or a guaranteed low-interest loan does now.

The grid, in the Albertsons case, would seem to be little more than a neutral conduit between power plants and refrigeration devices. Even a decade ago this impression would have been fairly accurate. As it turns out, however, all those smart meters shooting out wireless streams of data 24/7 are good for more than just helping the utility know that they should dispatch a lineman to your neighborhood before you call them up and tell them to. Within five years of the rollout, the data produced by smart meters was proving essential to the creation of predictive models of electricity use, minute by minute, as well as providing occasional real-time data about peaks and troughs in variable and distributed electricity production. And they were enabling real-time "demand-response," which is to say that they gave the utility the capacity to ask big electricity consumers to ramp down consumption as a means of balancing the grid. Rather than going offline and using diesel generators to provide backup power for a bit while the utility straightened things out, smart meters can be linked to efficient buildings that automatically deploy grid-scale conservation. At times this is accomplished by something as simple as dimming the lights. Negawatts, in other words, can now be ordered up by the utility and delivered by an Albertsons. Network enough of these power-savers into a flexible, smart piece of software, and you have your platform.

This demand-response capacity, called DR in the business, not only brings energy saved into the mix of resources available to grid operators by literally making conservation count, but it is another non-thing slowly taking grid governance by storm. And it doesn't stand alone. With computerization demand-response can be linked to other machines for making or saving power that we are currently building, willy-nilly, into the grid. Most of these are paid for by a mix of interested parties including investors, utility companies (which is to say, rate-payers), and state and federal subsidies. When enough of these scattered but existing resources are networked together it is possible to create what is called a virtual power plant. Not a plant for making virtual power, but a platform that connects everything available to it and gets it all to work *like* a power plant.

A virtual power plant can link, for example, a big coal-burning plant to a local military base's microgrid to three cogeneration plants to seven smaller natural gas combustion turbines to thirty-five hundred rooftop solar installations (three hundred of which also have deployable battery storage) to fourteen reliable, flexible, medium-scale negawatt producers to thirty thousand electric cars. It can then use the resources of each—generation, deployable efficiency, storage—to balance out demand with production capacity throughout the system by the millisecond.

Such interconnections between resources allows us to keep the *idea* of a power plant without our necessarily needing the power plant itself. A virtual power plant is thus primarily an organizational tool that uses information about electricity transmitted by electricity (digital smart meters most especially) to respond to the ebb and flow of power on the grid with a degree of timeliness and nuance that a human simply cannot match. We have always been too slow for electricity, but with smart meters, thousands upon thousands of distributed microsensors and the right computer programs, we can at long last do something about it.

The main issue with virtual power plants isn't bringing them to pass; they already exist. There are small versions running all over the place. The problem is getting all the necessary components into the mainstream (the cars most notably are still lacking, but smart appliances would make a big difference too), getting them all to speak the same language, and figuring out how to move through regulatory regimes and ownership blockades still in place from the twentieth century's far more proprietary and centralized grid.

Our first vision for a possible future of our grid, outlined in the last chapter, was of a material system unimaginable to us, yet being dreamed, tinkered, and built despite our incapacity to see exactly where it might all be going. Virtual power plants allow us a peek at a vision for the future of electrical infrastructure nearer at hand—a grid so jam-packed with computerization that it sparkles. This grid, however, would not look futuristic. Much like that suburban Albertsons, it would appear to the untrained eye to be identical to the grid we have. The revolution would not be one of form but of function.

The technology for this largely already exists, and the processes that might bring it to pass are already being put into place. In October 2015, the Supreme Court heard oral arguments in a case that pitted the Federal Energy Regulatory Commission (FERC, the folks charged with managing our grid nationally) against the Electric Power Supply Association (ERSA, which advocates for competition among power providers). At issue was precisely the question of whether a watt saved would be compensated at the same rate as a watt made. Unlike the California case, where the legislature narrowed the state's possibilities for grid reform by excluding the most troublesome elements of their current grid from their accounting, this case is being adjudicated in full view of the nation. And the stakes for how we make, use, and value electric power are also much higher.

If the capacity to integrate energy efficiency is to become central to the way public power is managed in this country and if virtual power plants are going to be among the means we use to network our resources—those that produce power as well as those that negate or reduce consumption—then a negawatt will need a stable value. It will need to become a currency whose worth everyone can agree upon so that it might be transacted without unreasonable risk.

The Federal Energy Regulatory Commission, the only regulatory body with a mandate to govern our electrical system, felt that a "commitment to reduce demand" should "be compensated the same amount as an equivalent commitment by generators to increase supply." This was not just about fairness, but also a means of making the grid a more integrative machine. Or, in their words, "paying demand response providers the full value of their contribution to the market would help overcome preexisting barriers to demand-response participation and increase the reliability and competitiveness of wholesale markets." Legislation can be used to wrench open rather than delimit who participates in our grid, whether as producers of power or of savings. FERC selected the more difficult path, but the one that will force an inclusivity—a technological as well as a financial problem—into the substrate of our grid.

The utilities, who are once again the ones struggling to maintain something like a viable revenue stream, begged to differ and sued FERC for jurisdictional imprudence. They argue "that real energy generated by real power plants should attract a higher price, in order to stimulate much-needed investment." Here again the battle's lines take a familiar form. Some people value stuff, real existing stuff, more than non-stuff, nothings, and zeros that can be added, counted, and made manifest. Others think that negative 1 should be seen as the equal and opposite of 1 and that we have a strong enough mathematical system already in place to account for, measure, and value what isn't there every bit as much as what is. The stakes for the grid of this argument is whether all the various people producing, saving, balancing, counting, and making a profit off of power will be tied into a common system or if only the big guys with their big power plants will matter. Regardless of what the court decides (as this book goes to press they are in deliberation), even a minor shift in price or mandate will produce major ripples through a near future system for making and distributing electric power.

If, in all of this, nothings can be given an agreed upon, transactional value then virtual power plants might help us integrate all the resources at our disposal, material and immaterial, legislative and corporate, collective and individual—such that the whole system runs more efficiently. In this way we might radically diminish how much power we are making and equally radically increase how much of this power will come from renewable sources. These are the first necessary, if small steps, toward engineering a changed energy landscape.

In this version of the future, we get to keep the big grid for a while longer and most everything familiar from it—wires, substations, long-distance AC and electricity markets. We probably even keep the utility companies by figuring out how to pay them for something other than how much power we consume. We could pay them for gadgets perhaps, or a basic line fee per connected meter, or as consultants to newly aggregating towns and newly organizing microgrids, or as innovators in the still turgid waters of our energy future. The only thing we lose with this new version of our grid is our reliance on central stations.

Big generation might stick around (big wind farms and big solar plants are proof of that) but it needn't. Computing has given us the capacity to tear the heart right out of the center of our grid. It doesn't need a heart. It can work with a million hearts, or a hundred million, scattered to the four winds and brought into balance by reactive, sensitive, ubiquitous software. Virtual power plants and their kin, the Energy Cloud, do something that microgrids and nanogrids threaten to undo. They keep the grid and its generic power for the people. All of us, together.

With good battery systems, which are just now reaching the level of ease and affordability that make them interesting to residential customers, Americans could, like the Germans, decide to defect entirely. And yet we aren't doing this. We may be happy to starve the system of cash by paying our money to some company that is not a utility, but even so we stay plugged in to the grid. I suspect this isn't just because of a communitarian spirit, though there is a bit of that in the mix. Rather, it is largely because "alternative energy" has been sold in America as a money-making scheme that renders each home solar installation a tiny factory for producing watts that are guaranteed by law to be purchased at market price. The seller—you or I—doesn't have to do a thing to insure this. No marketing, no sales pitches, no advertising campaigns, no haggling over the price. Going off the grid would mean losing this income, and it would also necessarily involve investing a lot more money in home storage systems and other small-power solutions for balancing the unwieldy electrical demands of a single customer. Getting off the grid would, in other words, be costly and complicated, just as it has always been. Insull sold Chicago on central station power to begin with, way back at the turn of the last century, on precisely these same terms. Sharing, when it comes to electricity, is simpler and more cost effective, than doing it for oneself.

The pragmatics of this simple truth are the elementary bond that keeps us together, but it's a weak bond rather than a strong one. It is only going to get easier to get off the grid, especially if a customer base funnels its frustrations with the current state of the grid toward technological, and business, developments that facilitate defection. It's complicated and expensive to get off the grid today, but in five years? In ten? If we want to keep America woven into one nation of equal opportunity then the grid, its technologies, and its wardens will need to pay more attention to what individuals expect of their infrastructure. This isn't always straightforward, since Americans in general don't see the grid for the enabling technology that it is, but symptomatically, our actions can be read and responded to.

One thing we certainly do want is a way to yank the cords out of our everyday experience of electricity. There are two developments on this front that will soon enter the mainstream. The first is to integrate wireless charging into most flat things, like shelves, countertops, tables, and lamp bases. This won't get rid of the outlet, as we'll still have to plug in the shelf, but it will render the plugs on most electric devices irrelevant. There is nothing radical about wireless charging: electric toothbrushes have been using this technology since the 1990s, and many of the forklifts in America spend their nights on flat pads that charge them wirelessly while their operators sleep. Within a couple of years we won't be plugging in any of our portable electronics though we will still likely have to toss, or park, them in the right spot.

In 2015, for example, Ikea rolled out a series of side tables, nightstands, and lamps with wireless charging pads so thoroughly integrated into them that they "simply blend in." Though these are not yet optimized—they won't charge Apple products, for example, leaving the world's 7 million iPhone users stuck with their ports, plugs, and outlets for a little while longer—they do clearly point toward a future in which outlets move us less and cords, adapters, and power strips are reduced, at long last, to nothing.

The second, and slightly more anxiety-producing, possibility for those worried about the effects of ambient electromagnetic radiation on their ball-shaped organs is the wireless distribution of electricity. The desire for wireless transmission is as old as the grid itself, and the capacity to effect this transfer of power through thin air over short distances has been viable for just as long. Nikola Tesla, the quirky Serbian-born inventor who brought us alternating current in the 1880s, also, in 1893, lit three bulbs wirelessly from a hundred feet away as a part of the Columbian Exposition in Chicago—a World's

Fair devoted to electric lighting in all its multitudinous, miraculous variations.

Though the claim that Tesla lit even more bulbs (rumor says a hundred) even more spectacularly from the ambient electricity produced by lightning strikes up to twenty-five miles away appears to be spurious (invented, apparently, by his biographer), what is sure is that Tesla could, just as we can today, move a viable electric current across a modest slice of air. He was equally devoted, later in his life, to the prospect of long-distance remote power projects that used the earth "literally alive with electric vibrations" as a giant conductor and tethered balloons, thirty thousand feet up in the atmosphere, to transmit electricity at billions of volts around the world. His hope was to make possible both universal wireless illumination and instantaneous, wireless communication between places as far distant as London and New York. He is quoted as saying, in support of this undertaking:

"It will soon be possible, for instance, for a business man in New York to dictate instructions and have them appear instantly in London or elsewhere. He will be able to call up from his desk and talk with any telephone subscriber in the world. It will only be necessary to carry an inexpensive instrument not bigger than a watch, which will enable its bearer to hear anywhere on sea or land for distances of thousands of miles. One may listen or transmit speech or song to the uttermost parts of the world."

This was 1909. A century later we could do all these things. The only thing we couldn't do was wirelessly transmitting the electric power necessary to keep this "device not bigger than a watch" running. The telecommunications revolution that Tesla understood to be part and parcel of the mass, wireless, electrification of all the world grew to be a dream split in two. Everyone has cell phones, even in places with nothing like a reliable electric grid. But very little of the world has the wireless communication of electricity to power them.

Be that as it may, Tesla's dream still seethes below the surface of our wished-for way of interacting with electrical infrastructure, even more so now that wirelessesness has become so commonplace in so many other domains of daily life.

Long-distance wireless transmission of electricity remains something of a fantasy, Tesla's own efforts were neither better nor worse than anyone else's on this front. It turns out, however, that short-distance transmission is a relatively straightforward project, according to Croatian-born Martin Soljačič, a professor at MIT.

Soljačič has figured out a system that, though fairly different in its physics from Tesla's, has much the same effect—the wireless transmission of enough electricity across empty air to light a bulb. Soljačič even put a wooden board between his bulb and the transmitter and it still emitted a lovely glow. What is remarkable about Soljačič's system is that it doesn't fill the air with electromagnetic radiation, as Tesla's system did, but rather uses magnetic resonance, which essentially only allows the electricity sent to power a device specially "tuned" to receive it. This use of magnetic fields targeting appropriately chipped electronics helps these devices to sidestep known problems with wireless transmission via electric fields, which have a long range but poor aim, enveloping everything in a thick, invisible haze of electromagnetic radiation.

Since Soljačič worked out the basic physics of point-to-point wireless transmission in 2007, the start-up market has exploded with products, marketing alliances, and live demos. Competitors are merging or eating each other as the struggle to bring a viable wireless charger to market heats up. We don't see much of this yet, as the winners and losers are still being worked out behind the scenes, but we will soon. Says Brian Krzanich, the CEO of Intel (which is pushing this technology hard): "Imagine a world where you can charge your devices wherever you are. That's the world I want to live in." This is a pretty safe thing for him to say after a massive survey conducted by

Intel across forty-five countries revealed that four of the most common complaints people had about their computing devices had to do with the cords. Given that we can already transmit power wirelessly across about three feet of thin air with 90 percent efficiency, in time we will.

Taken together it can be said with some certainty that we'd like the grid to move us less, to be less polluting, more adaptive, and more reliable. We'd like systems change to be more about responding to the powers to come and less committed to maintaining the powers that be. We'd like some control over how our power is made and also some legible way to understand how it is used. Despite all of this we'd also prefer a grid—an electrical system—in common.

There is no reason to believe that the spirit of the time, orienting all of us toward things which do not exist, will abate in the near future. The odds rather predict the inverse. More wireless, more renewable, more portable, more integrated, and thus more invisible technologies that we move, rather than that move (or worse, root) us. Neither is there reason to believe that despite a strong cultural preference for the invisible, the immaterial, and the hidden that we will abandon the idea of a grid in common. What remains to be seen is the precise shape this infrastructure will take once these two biases have been twisted into one. My vote is for a beautiful one, minimally invasive, that shines rather than glowers and that is wrought into the leitmotif of the century we are only just now stepping into in earnest.

Much has been made of errors and faults as signs of beauty in Japanese aesthetics. In contrast to Western notions of beauty that have everything to do with symmetry, balance, and perfection, in Japan flaws and irregularities are highlighted. And where they might otherwise be overlooked—a cracked pot perfectly repaired—they are brought deliberately to the fore and made to catch the eye. In the case of the broken pot, for example, gold or bright red lacquer might be rubbed into the seams when it is mended. So rude is this interruption to the form of the thing that one hardly sees the original pot anymore. The trace—a sparkling vein; a rivulet in bloodred—these outshine the fact that the object has regained integrity. The fact of having been broken is what matters. The history of a thing—its difficult life—is made to be a part of its attraction.

In the United States, that same pot, clumsily dropped and shattered into shards, might be carefully repaired, the owner (epoxy in hand) taking great care to find all the bits and to reunite them into a seamless, usable whole. More likely, though, the broken ceramic would be blithely swept up and deposited without ceremony in the trash can, and another bowl would be bought on the morrow, or perhaps even ordered online for next-day delivery.

Americans don't like dealing with remnants or with imperfect things. We don't want to see that what now serves us perfectly well was once broken or damaged. We want youth and vigor, not old age and a storied life. Repair is not a cultural value. Replacement is.

And yet, if we are to maintain a grid with a national, or even half-national, span, we will need to change our minds about what constitutes a "good" thing. A good thing might very well be an old thing with veins of gold pressed into all the cracks. It might be a mended thing with the mends themselves constituting the most precious element of the whole.

If we want to keep a grid for all, we might be wise to mend our grid like a Japanese pot. The most valuable bits, the golden threads, the tiny machines, we rub into all the seams—the glue would matter most. In the case of the grid, this glue isn't real gold but rather millions of tiny machines—microprocessors—that when working together have the reactive capacity to make decisions. The system as a whole would be given the approximate intelligence of a tick, which has only four or five basic capacities: to climb, to smell food, to drop, to seek warmth, to eat. As Ronald Bogue points out, "The tick's milieu is a closed world of elements, outside of which nothing else exists"; nevertheless it's still a whole lot smarter, more capable, and more flexible than a hunk of rebar or a puddle of tar. This tick-bright grid would react and communicate well enough that its complexity might become its strength. And because it will have a microprocessor pushed into every crack it will be able to take anything we can throw at it—variable generation, distributed generation, small power, big power, negawatts, nanogrids, mobile storage, weird weather—and integrate these into a self-balancing, highly reliable system. Such a grid would be something like a national (or half-national, or regional, or whatever size we choose to make it) computer in which all that is old, rusty, and broken is healed, and held together, by the densest network of intelligent agents anywhere on the planet.

Not for long, of course, as ubiquitous computing is coming to, well . . . everywhere.

The grid that virtual power plants, and their ilk, imagine making possible, the grid their promoters are trying to build with elbow grease, a golden tongue, and difficult-to-extract handshakes across the front lines of old battlefields, might well be the first step toward a larger, more ambitious project: a self-healing, processor-dense, "intelligent" grid. One that heralds a sort of immersive technology our children's children will take for granted in much the same way that Edison's first mile of wires and bulbs laid into the muck-thick streets of Manhattan a century and a half ago heralded the present age of "ubiquitous" electricity.

In order for this new grid to come to pass, its architects, its dreamers, its schemers, its malcontents, and its make-a-buck quicksters will all need to find their way into this vision of future technology, not as an ideal to be realized but as a pragmatic route to the best possible end. Wealth, of course. Growth, of course. Excess, of course. But also a chance for something more than these standard business goals of standardized businesses.

It's not hard to have a less polluting, less irritating, and more reliable electrical system than the one we have. What is hard is figuring out how to bring big dreams, smart inventions, and popular will together with the entrenched interests of the powers that currently govern, own, and make a profit from our grid. The easiest way to do this is to force upon them, at every possible opportunity, a radical openness to variety—to avoid California's path, with its eyes closed to small power producers; to avoid England's path, where a shift away from big coal-burning plants has resulted in a grid-scale reliance on diesel generators; and to avoid Germany's path, where the exploited (in their case, the companies) have walked away from public power, taking their poolable resources with them.

The future we want is one in which difficult things are integrated, even when this is a more troublesome route than excluding them. Let's take them all, every variation on the theme of "grid"; let's consider them all, every form of belief about how electric power should be made and used (or not made and not used); and let's integrate them all in a way that does the least planetary damage over the long term.

A wise grid is not just a smart grid ramped up a little bit. Wisdom is not the product of added computational power (though that helps). It's a mode of systems reform capable of hearing what people say, noticing what they do, and premising thoughtful action upon both. Wisdom, when speaking of the grid now, is about helping people accomplish well what they have already begun. It is about following vectors of desire and modes of action forward and then building these in at the level of the infrastructure itself.

I don't know what America's grid will look like in thirty years. Nobody does, precisely because in America today grid reform is a groundswell, it is barely organized; it has no single valance and no political party touting a particular path. There are crusaders, but they are the sort that go door to door and ask their neighbors to join in a solar co-op. There are warriors, but they are the sort who petition their local government to allow anyone to opt out of a smart meter. There are visionaries, but they are thinking about how the seven homes that share a transformer might organize themselves into a collective of diverse enough generation and significant enough storage to keep the lights on the next time the big grid blacks out. There are inventors, but odds are that they're going to tell you a story about self-inflating hydrogen balloons in the desert. The future is a crapshoot. If we are smart enough, it might also be a chance to capture the cutting edge of technological innovation and cultural imagination and concretize it in the grid itself. All the visions of ubiquitous technology, sentient cities, chips everywhere could well take their alpha form in the electric grid. It is, after all, as Nicola Tesla pointed out, not only a system for powering the world but also essential to the lines of communication that weave our economies, our labor, and our imaginations together. If we are going to bring the Internet of Things into our daily lives, then why not start with the biggest thing of all? The grid, tick-bright and aglow with promise.

Afterword

Contemplating Death in the Afternoon

As I write this, the power is out. It's below freezing outside, though it's midafternoon on a sunny day in early spring. I have a couple of hours of battery power left in my computer. I was using it for most of the morning without having plugged it in, though there was an outlet less than a cord's length away. I am kicking myself for that now. Despite my computer's pending demise, it's difficult to think of this power outage as a "blackout"—the house is flooded with sunlight. There is no hint of darkness. For the moment, at least, everything is fine, except perhaps a slight, nagging irritation that my singular task for the latter half of the day—writing approximately four thousand overdue e-mail replies—has just become impossible. No power, of course, means that there is no Internet, either.

Despite this now unavoidable failure of e-etiquette, I am mostly worried about the cold. It will be fine indoors for a couple of hours, but I have a child to feed and house at the end of the school day, and tonight's temperatures are heading to down into the teens.

Our heat is electric, our stove is electric, our hot water heater electric; I may be writing a book about the grid, but I, like almost everyone else, also have to live with it. It is my intimate. The grid's good days and its bad days structure my life, my capacity to work, my reputation, my ability to care for my family. Its material quirks form the undulations of my worry. And tonight, as the blackout becomes real, I'll wrap myself and the thickest of our blankets around my son's small body. As he sleeps, I'll read by candlelight a book about men slowly dying of thirst in the middle of the Pacific Ocean (not a happy story) and I'll wonder if this is us too.

We live in a sea of electricity. Our grid, for now, is the means we use to bring this electricity home. It warms and nourishes us. It makes life in extreme environments like Fargo and Phoenix livable, even comfortable. Despite being surrounded by this abundance of power, if we cannot make the means to this electricity, more reliably our own, we, too, could die of "thirst." It does not matter how much water is in the sea, or how much electricity we generate, if that water is not drinkable and that electricity is not delivered.

I have written this book as an optimist, the dreams of a better future are so strong, so palpable, so motivating. And then the power goes out and I, like almost everyone else, don't have a backup system in place. No solar panels, no bank of batteries, no diesel generator, no gas heater, though I do have a rotary phone (this much I learned from the Straws). It's the second decade of the twenty-first century, in one of the richest countries in the world. It is time, I think, for us to do better than this.

Acknowledgments

In the early days of this project I spent a great deal of time interviewing experts in every domain of grid governance, innovation, and upkeep, going to conferences and public talks, and generally learning from the know-how of those who have spent their lives in the midst of this inelegant if remarkable machine. I cannot name you all here, but let it be known that your wisdom was invaluable to my own understanding. Special thanks to the Huntington Library for their generous support of research in their archives, and in the papers of Southern California Edison most particularly.

The grid is complex, impossibly so; despite the wisest advice from the sagest minds, I have most certainly gotten things wrong, though I hope only little things. These errors are my own.

Special thanks go to my stupendous agent, Susan Rabiner, without whom this anthropologist's dream of telling the story of a vast and unwieldy system differently would never have made its way into print. And also to my editors, of whom there were many, each and every one having left his or her mark in the form and unfolding of the story herein. It is a much better work for their collected efforts. Last, it is said that in every relationship there is a gardener and a flower. For all the years I have spent picking the grid apart and finding a way to give it over into words, my husband, Julien Weiss, has been that gardener, and I am intensely grateful.

The transformation that the grid has undergone during the six years I spent researching and writing this book are almost inconceivable in their scope. I can only hope nothing slows down (or breaks down) as we abandon the grinding route toward reforming our energy infrastructure and opt for the full-on sprint.

Notes

INTRODUCTION

still goes into cars: "To achieve the emissions goals," the *New York Times* tells its readers, "the entire economy, including transportation, needs to be electrified as much as possible. That might mean cars running on batteries, but it could also mean cars running on hydrogen, created by using nighttime electricity from nuclear reactors or wind turbines to split water molecules. Either way, the implication is that the internal-combustion engine that has powered cars since the nineteenth century is a technological dead end in the twenty-first. So countries like the United States that are spending a lot of effort trying to make gasoline cars more efficient may be going down a blind alley." Justin Gillis, "A Path for Climate Change, Beyond Paris," *New York Times*, December 1, 2015, http://www.nytimes.com/interactive/2015/12/01/science/19781608196104406576.embedded.html. See also Justin Gillis, "Short Answers to Hard Questions About Climate Change," *New York Times*, November 28, 2015, http://www.nytimes.com/interactive/2015/11/28/science/what-is-climate-change.html.

are remarkably oblivious to it: So deemed by the National Academy of Engineering in its "Greatest Engineering Achievements of the Twentieth Century," www.greatachievements.org.

(its infrastructure for itself): Like Texas, which has preserved an independent electrical infrastructure unlinked to those of its neighbors, Quebec also has an electric grid all its own. Located in the Canadian province most likely to secede, it's the best in North America.

an American power plant: This statistic is taken from Campbell 2012, as cited in "Economic Benefits of Increasing Grid Resilience to Weather Outages"

(Washington, D.C.: Executive Office of the President, August 2013), 7, as well as "Frequently Asked Questions: How Old Are U.S. Nuclear Power Plants and When Was the Last One Built?" U.S. Energy Information Administration, February 20, 2015, https://www.eia.gov/tools/faqs/faq.cfm?id=228&t=21.

from 15 in 2001 to 78 in 2007 to 307 in 2011: In the first six months of 2014 alone, there were 130 significant outages. Compare this to the previous fourteen years' worth of outages:

174 in 2013	123 in 2010	78 in 2007	93 in 2004	15 in 2001
196 in 2012	97 in 2009	91 in 2006	61 in 2003	30 in 2000
307 in 2011	149 in 2008	85 in 2005	23 in 2002	

Jordan Wirfs-Brock, "Data: Explore 15 Years of Power Outages," *Inside Energy*, August 28, 2014, http://insideenergy.org/2014/08/18/data-explore-15-years-of-power-outages/. See also Christine Hertzog, "Why Climate Change Will Force a Power Grid Makeover," *GreenBiz*, August 23, 2013, http://www.greenbiz.com/blog/2013/08/23/climate-change-power-grid-makeoever.

679 between 2003 and 2012: "Economic Benefits of Increasing Grid Resilience to Weather Outages" (2013).

Germany at 15: Craig Morris, "German Grid Most Reliable in Europe," *Renewables International*, July 19, 2011, http://www.renewablesinternational.net/german-grid-most-reliable-in-europe/150/407/31462. For Korea and Japan, see Andy Bae, "Blackout in Seoul," *Navigant Research*, October 4, 2011, http://www.navigantresearch.com/blog/blackout-in-seoul.

Because the United States doesn't aggregate its statistics nationally when it comes to annual outage minutes, the estimates vary wildly, from about 1.5 hours (for the Midwest) to 9 hours. Meagan Clark, "Aging U.S. Power Grid Blacks Out More Than Any Other Developed Nation," *International Business Times*, July 17, 2014, http://www.ibtimes.com/aging-us-power-grid-blacks-out-more-any-other-developed-nation-1631086; and "Power Blackout Risks: Risk Management Options," *Emerging Risk Initiative—Position Paper* (CRO Forum, November 2011), https://www.allianz.com/v_1339677769000/media/responsibility/documents/position_paper_power_blackout_risks.pdf.

it's less than ten minutes and shrinking: Peter Asmus, "How Microgrids Improve Grid Reliability and City Resilience," *GreenBiz*, December 5, 2012, http://www.greenbiz.com/blog/2012/12/05/how-microgrids-build-resiliency-extreme-weather.

"without power for two or more hours": Massoud Amin, "Why We Need Stronger, Smarter Electrical Grids," *GreenBiz*, July 29, 2014, http://www.greenbiz.com/blog/2014/07/30/why-we-need-stronger-smarter-electrical-grids.

squirrels most especially: "Longmont Power & Communications, which serves 35,000 customers north of Denver, says that more than 90 percent of its significant outages are caused by squirrels." Barbara Carton, "Fried Squirrel Fails to Find Favor With Public Utilities," *Wall Street Journal*, February 4, 2003, http://www.wsj.com/articles/SB10443096593731245584.

burn and then have no more: "Stock resources are those resources whose 'total physical quantity does not increase significantly with time . . . each rate of use diminishes some future rate.' . . . Flow resources are those resources where of different units become available for use at different intervals . . . the present flow does not diminish future flow, and it is possible to maintain use indefinitely provided the flow continues" Maurice Kelso, "The Stock Resource Value of Water," *Journal of Farm Economics* 43, no. 5 (December 1961): 1112.

of the power on our grid: 38.5 percent coal, 27.3 percent natural gas, and 0.7 percent petroleum. From "2014 Renewable Energy Data Book" (National Renewable Energy Laboratory (NREL) of the U.S. Department of Energy, November 2015), http://www.nrel.gov/docs/fy16osti/64720.pdf, 10. See also "Frequently Asked Questions: What Is U.S. Electricity Generation by Energy Source?" U.S. Energy Information Administration, March 31, 2015, https://www.eia.gov/tools/faqs/faq.cfm?id=427&t=3.

solar power are most effectively produced: It is arguable that wind-poor and sun-poor locations closer to where people live and use electricity may well have sufficiently robust (if middlingly) resources to meet demand. Personal conversation with Amory Lovins, May 2016.

to replace the retiring fleet: Karl Mathiesen, "Gas Surges Ahead of Coal in U.S. Power Generation," *Guardian,* July 14, 2015, http://www.theguardian.com/environment/2015/jul/14/gas-surges-ahead-of-coal-in-us-power-generation.

for the climate than coal: There is some argument over the accuracy of these numbers. According to Michael Obeiter, a senior associate in the World Resources Institute's climate program, "There's not a lot of good data for methane leakage nationwide, but it's probably less than 3.2 percent. The EPA GHG [greenhouse gas] Inventory estimates roughly 1.2 percent or so, but many have taken issue with their methodology." See also Ramón A. Alvarez et al., "Greater Focus Needed on Methane Leakage from Natural Gas Infrastructure," *Proceedings of the National Academy of Sciences* 109, no. 17 (2012): 6435–40. Gal Luft, an adviser to the United States Energy Security Council, points out: "The greenhouse effect of methane is about 17 times that of carbon dioxide. However, we know that if the overflow ratio exceeds 2 percent methane, the greenhouse effect of shale gas mining may be more severe than coal. Current measurements have recorded 2.5 percent, 3 percent and 10 percent. Currently, we are still unclear of the exact measurements; many basic problems of shale gas development have yet to be solved." Wang Erde Wei Wei, "Nuclear, Not Fracking, Is the Answer to China's Future Energy Needs," *China Dialogue*, July 25, 2013, https://www.chinadialogue

.net/article/show/single/en/6228-Nuclear-not-fracking-is-the-answer-to-China-s-future-energy-needs-.

nuclear power development: Wei Wei (2013).

"will need to be replaced by new plants": James E. Rogers, in his foreword to Peter Fox-Penner's *Smart Power Anniversary Edition: Climate Change, the Smart Grid, and the Future of Electric Utilities* (Washington, D.C.: Island Press, 2014), xv.

twenty times as much solar: These numbers are from an August 2015 speech by Barack Obama introducing his "Clean Power Plan," the aim of which is to reduce carbon emissions by 2030 to 32 percent lower than they were in 2005. Power plants are responsible for about one third of current carbon emissions nationally, and thus the plan should have a noteworthy impact on plant retirements, upgrades, and choices of replacement technology. Since the plan also addresses methane emissions, a significant portion of which occur during the extraction of natural gas, it will also affect the cost of using natural gas as a replacement for coal or nuclear.

about 7 percent overall: To be precise, nonhydro renewable energy sources amount to 6.76 percent. Wind accounts for 4.42 percent, biomass wood for 1.04 percent, biomass waste for 0.52 percent, geothermal for 0.39 percent, and 0.39 percent is from solar. "How Much U.S. Electricity Is Generated from Renewable Energy?" U.S. Energy Information Administration, June 12, 2015, http://www.eia.gov/energy_in_brief/article/renewable_electricity.cfm.

it's a stunning 30 percent: "U.S. Wind Energy State Facts," American Wind Energy Association, accessed September 15, 2015, http://www.awea.org/resources/statefactsheets.aspx?itemnumber=890.

dropped to negative 64¢: Eric Wieser, "ERCOT Sets Wind Generation Output Record Sunday, Real-Time Power Prices Move Negative," *Platts McGraw Hill Financial*, September 14, 2015, http://www.platts.com/latest-news/electric-power/washington/ercot-sets-wind-generation-output-record-sunday-26208539.

has more than doubled: It was 3 percent in 2009, according to Dr. Stephen Chu, the Nobel Prize–winning physicist and former U.S. secretary of energy who spoke at Grid Week in Washington, D.C. (see chapter 1).

Maine is aiming for 40 percent: Jocelyn Durkay. "State Renewable Portfolio Standards and Goals" *National Conference of State Legislatures*. October 14, 2015, http://www.ncsl.org/research/energy/renewable-portfolio-standards.aspx.

Hawaii is aiming for 100 percent: Ari Phillips, "Hawaii Aims for 100 Percent Renewable Energy by 2040," *Renew Economy*, March 13, 2015, http://reneweconomy.com.au/2015/hawaii-aims-for-100-renewable-energy-by-2040.

"month's worth of purchases": W. Kempton and L. Montgomery, "Folk Quantification of Energy," *Energy* 7(10) 1982: 817–27.

"benefits of their actions": Kathryn Janda, "Buildings Don't Use Energy: People Do," *Architectural Science Review* 54, 2011: 15–22.

would have happily left behind: We can see this history of uneven development in Vermont, where in 1920 only 10 percent of the farms in that state had electricity; by the start of the Great Depression, the percentage had risen to 1 in 3. In the decade after the passage of the Rural Electrification Act (1936), most of the state's rural residents had electrical power, in their milking parlors if not yet in their homes. And in 1963—almost a hundred years after America's first electric light—its final two towns, Granby and Victory, Vermont, got their wires and became a part of our grid. This is just one example, but it's an important one: electrification was not an all-at-once affair; in some spots its history is far thinner than in others. From a radio commentary by Vic Henningsen on Vermont Public Radio, August 27, 2015, http://digital.vpr.net/post/henningsen-statewide-service.

"works in practice, but not in theory": Alexandra von Meier, "Electronic Power Systems," September 17, 2010, Public Lecture i4 Energy Systems, University of California, Berkeley.

America, didn't know how to change: In part, according to Richard Hirsh, utility hiring practices selected for the risk-averse, noncreative, bottom of the engineering barrel. This led to conservative reactions to new problems. Richard Hirsh, *Technology and Transformation in the American Electric Utility Industry* (Cambridge: Cambridge University Press, 1989).

(utility since the Depression): The California blackouts in 2000–2001 were not caused by degraded infrastructure but by a perfect storm of bad legislation, criminal profit-mongering, and poorly designed infrastructure (see chapter 4).

"emissions of greenhouse gasses": "Economic Benefits of Increasing Grid Resilience to Weather Outages" (2013), 3.

constructing eighty-three new microgrids: "Microgrid Deployment Tracker 4Q15," *Navigant Research*, accessed December 15, 2015, https://www.navigantresearch.com/research/microgrid-deployment-tracker-4q15.

CHAPTER 1 : The Way of the Wind

power 4.5 million households annually: The Foundation for Water & Energy Education estimates that there is the potential to produce 100,000 MW in the Gorge, though this would require blanketing much of the available space with turbines. "Wind Farms & Northwest Energy Needs," *Foundation for Water & Energy Education*, accessed December 1, 2013, http://fwee.org/nw-hydro-tours/how-wind-turbines-generate-electricity/wind-farms-northwest-energy/.

any fuel will do: Dried cow dung, like oil, is a negligible source of electricity production in the United States.

over the sun just as quickly: The problem with solar power, and to a lesser degree wind power, isn't only these dramatic instances of total cloud cover or breeze-to-stillness, but also that generation is actually very jiggly. Solar panel output shifts five or six times a minute, and a field of solar panels doing this in sync is difficult for the grid's automation to balance.

a full twenty-four hours to turn either up or down: Eric Hittinger, J. F. Whitacre, and Jay Apt, "Compensating for Wind Variability Using Co-Located Natural Gas Generation and Energy Storage," *Carnegie Mellon Electricity Industry Center Working Paper* CEIC-10–01 (December 2010).

can go to Mars: 55.8 million miles in five minutes. Mars's distance varies from Earth; sometimes it's closer than this, sometimes farther away. But it wouldn't be wrong to say that on a good day, under perfect transmission conditions, we could send a current from Indiana to the red planet in about five minutes.

before its sixty seconds are up: The 2003 blackout of the Eastern Seaboard, the third-largest blackout in this history of the planet, essentially happened in thirty of the seconds between 4:11 and 4:12 p.m. There were a lot of problems leading up to these very bad thirty seconds, but the imbalance that tipped over into darkness propagated in about the time it takes to draw in a breath.

meltwater are more than sufficient: On an average blustery day, the wind power in the Gorge produces enough power for three times as many people as live in Oregon.

"Northwest has ever experienced": Ted Sickinger, "Too Much of a Good Thing: Growth in Wind Power Makes Life Difficult for Grid Managers," *Oregonian*, July 17, 2010, http://www.oregonlive.com/business/index.ssf/2010/07/too_much_of_a_good_thing_growt.html.

let the water out through spillways: There is also pumped storage on the Columbia, and excepting years of extreme drought this is all also full in May.

from dinner plates to Grandma's memory bin: Full disclosure: my father, Bill Bakke, did this (made it illegal for the spillways to operate in the spring). He is not well loved by many in the electricity industry, but we as a nation still eat a lot of Pacific salmon, and without this law the palates of a generation would already be otherwise.

ferocious speeds in the onslaught of wild air: Most newer wind turbines can be turned down via adjustments to pitch and yaw, this was not so much the case in 2010. Such a basic technological improvement is one of the many small things under way that make the integration of variable renewables into our national electricity system increasingly plausible.

care for the grid: In 2013, 42 percent of Spain's electricity demand was covered by renewables, 21.2 percent wind, 3.1 percent solar photovoltaic, 1.7 percent solar thermoelectric, 2 percent renewable thermal, and 14.2 percent hydroelectric. "Corporate Sustainability Report 2013: 4. Committed to Security of Supply, Efficient Management and Innovation" (Red Eléctrica Corporación, May 7, 2014), http://www.ree.es/sites/default/files/02_NUESTRA_GESTION /Documentos/memoria-2013/English/RC/RC13_07_en.pdf. Despite this centralization, Spain's grid suffers from similar kinds of problems as Iberdrola faces in Oregon. For example, around Easter 2013, Spain saw "extremely low demand, high production of hydroelectricity with dumping in some basins, and a high producible wind power . . . Given this scenario, to ensure system security it was necessary to give orders to reduce production to a level not seen to date. These reductions affected, among others, nuclear production an exceptional fact [sic] and unprecedented since 1997." "Corporate Sustainability Report 2013" (2014), 53.

machines are turned on and running: There were 2,760 turbines in 2011, so three thousand in 2014 is an educated guess. Miriam Raftery, "A Walk Through the Wind Farm with Iberdrola," *East County Magazine*, April 2012. http:// www.eastcountymagazine.org/walk-through-wind-farm-iberdrola.

electricity these machines make: Some of this has to do with money, but a lot of it has to do with the way the utilities are run and managed. Historically they have been given a form—the so-called "natural monopoly"—that is slow to change, innovate, or cede power to anyone else. This is history and the principal topic of chapters 3 and 4.

line out to the site: The largest wind farm in Texas as of 2015 was the Roscoe Wind Farm, owned and operated by E.ON Climate and Renewables, which began operating in 2009. Eileen O'Grady, "E.ON Completes World's Largest Wind Farm in Texas," *Reuters*, October 1, 2009, http://www.reuters. com/article/2009/10/01/wind-texas-idUSN3023624320091001#1Uc6qUxR CSELa3Zr.97.

"The turbines installed at the farm range in between 350ft and 415ft tall, and stand 900ft apart. Out of the total number of turbines employed, 209 were the Mitsubishi 1000A model, with a rated output of 1.0MW." From "Roscoe Wind Farm," *Power Technology*, accessed November 8, 2015, http://www.power-technology.com/projects/roscoe-wind-farm/.

1.5 MW GE model: "The widely used GE 1.5-megawatt model, for example, consists of 116-ft blades atop a 212-ft tower for a total height of 328 feet. The blades sweep a vertical airspace of just under an acre. The 1.8-megawatt Vestas V90 from Denmark is also common. Its 148-ft blades (sweeping more than 1.5 acres) are on a 262-ft tower, totaling 410 feet. Another model becoming more common in the U.S. is the 2-megawatt Gamesa G87 from Spain, which sports 143-ft blades (just under 1.5 acres) on a 256-ft tower, totaling 399 feet." "FAQ:

Output from Industrial Wind Power," *National Wind Watch*, accessed November 8, 2015, https://www.wind-watch.org/faq-output.php.

"The average nameplate capacity of newly installed wind turbines in the United States in 2014 was 1.9 MW, up 172 percent since 1998–1999." From "2014 Wind Technologies Market Report" (U.S. Department of Energy, August2015),http://energy.gov/sites/prod/files/2015/08/f25/2014-Wind-Technologies-Market-Report-8.7.pdf.

"done right it's a huge opportunity": These quotes are from the Sickinger (2010) article. Manizer and I mostly talked about domestic hot water heaters and how to make interregional power swaps like the Western Doughnut the norm. These issues are addressed in chapters 8 and 9.

from renewable resources in 2012: A negligibly higher 13 percent in 2014. "How Much U.S. Electricity Is Generated from Renewable Energy?" (2015).

as a whole in 2014 was 6.76 percent: "How Much U.S. Electricity Is Generated from Renewable Energy?" (2015).

Dakotas, Iowa, and West Texas: Thomas P. Hughes, *Networks of Power: Electrification in Western Society* (Baltimore: Johns Hopkins University Press, 1983).

75 percent of all: "U.S. Wind Energy State Facts," American Wind Energy Association, accessed September 15, 2015, http://www.awea.org/resources /statefactsheets.aspx?itemnumber=890.

a 3,000 percent increase in a single year: These numbers are from the *2012 Renewable Energy Data Book* (National Renewable Energy Laboratory of the U.S. Department of Energy, October 2013), http://www.nrel.gov/docs /fy14osti/60197.pdf.

"there will be blackouts": Coral Davenport, "A Challenge from Climate Change Regulations," *New York Times*, April 22, 2015, http://www.nytimes .com/2015/04/23/business/energy-environment/a-challenge-from-climate-change-regulations.html.

"grid designed for the previous century": Evan Halper, "Power Struggle: Green Energy versus a Grid That's Not Ready," *Los Angeles Times*, December 2, 2013, http://articles.latimes.com/2013/dec/02/nation/la-na-grid-renewables-20131203.

CHAPTER 2: How the Grid Got Its Wires

during the previous weeks: I use the term "electrocution" here, but in the early days of electricity. "No standard words had yet been adopted for killing or death by electricity. Ones pondered by the *New York Times* included electromort, thanelectrize, celectricise, electricide, electropoenize, fulmen, voltacus, and electrocution." Nicholas Rudduck, "Life and Death by Electricity in 1890: The Transfiguration of William Kemmler," *Journal of American Culture* 21, no. 4 (1998): 86, note 8.

without being properly understood: On August 1, 1890, William Kemmler, a convicted murderer and inveterate drunk, was put to the chair and slowly roasted to death over a period of about eight minutes. Despite Edison's assurances (for he had designed and built the chair) that Kemmler's would be a swift, humane, and painless death, and despite the fact that the chair had been tested and retested and electricity of varying voltages had been used to efficiently kill all manner of things, from stray dogs to a retired circus elephant, Kemmler did not go out as planned. It was not his size that was the problem; Kemmler was a thin man, petite by today's standards. Nor was it a lack of sufficient voltage on the coal-fed, steam-powered 1,680-volt dynamo used to power the chair. The problem was that the wire connecting the chair in the Auburn prison to the dynamo in its basement was also being used that day to light thirty-six bulbs strung in parallel, which collectively siphoned off about a thousand volts, leaving a mere trickle of electrical capacity for the chair. What remained was enough to kill Kemmler eventually, but certainly not enough to kill him fast. It was a highly publicized horror that effectively ended Thomas Edison's career.

its ineffable physics: Gérard Borvon, *Histoire de L'électricité: De L'ambre à L'électron* (Paris: Vuibert, 2009), 1.

something like an instant: The first working telegraphs appeared in the 1830s, while the 1850s to 1870s saw the advent of intermittently functional transatlantic telegraphy.

displaced a less effective technology: Or as Isaac Asimov once said, "No steam engine or internal combustion engine, however powerful or however perfect, could run a television set (in the absence of electricity) with the direct simplicity electricity makes available to us." From a funny little pamphlet published by the U.S. Atomic Energy Commission: Isaac Asimov, "Electricity and Man" (United States Atomic Energy Commission Office of Information Services, 1972), http://www.osti.gov/includes/opennet/includes /Understanding%20the%20Atom/Electricity%20and%20Man.pdf, 19.

with it remotely fueled electric lighting: The power plant built at Niagara Falls was turned on in 1895, though it did not begin to transmit power to Buffalo until 1896.

always secondary to the story: The first "modern" dynamo for use in industry was invented independently by three different men in 1866; though Faraday gets the true credit in the early 1830s for a machine that makes electricity, his design was not a precursor of the next-generation machines, even though his ideas were essential to these.

by incandescent bulbs: Here Richard Moran was quoting a reporter from the *New York Times* in his book *Executioner's Current: Thomas Edison, George*

Westinghouse, and the Invention of the Electric Chair (New York: Vintage Books, 2002), 45.

but it was a grid: James C. Williams, *Energy and the Making of Modern California* (Akron, OH: University of Akron Press, 1997), 170.

3,000-candlepower arc lamps: 3,000 candlepower is a imprecise way to gesture toward the light three thousand candles would emit if they were all in the exact same spot. Though there is no good way to convert from candlepower to the brightness measure we are most used to—that of a 100-watt bulb—as a rough estimate, one 60-watt electric bulb generates the light of approximately a hundred candles. A 3,000-candlepower arc lamp would thus generate the light of something like an 1,800-watt bulb. Jon Henley, "Life Before Artificial Light," *Guardian*, October 31, 2009, http://www.theguardian .com/lifeandstyle/2009/oct/31/life-before-artificial-light.

to extraction sites: Williams (1997), 172.

bulbs strung in parallel: The 52 bulbs Edison had turned on in the *New York Times* editorial office were in addition to the 106 he had strung up in J. P. Morgan's office on Wall Street. Munson (2005), 18.

"90 percent of his labor": Munson (2005), 10. Original quote from Paul Israel, *Edison: A Life of Invention* (New York: John Wiley & Sons, 1998).

they buzzed disagreeably: Michelle Legro, "The Age of Edison: Radical Invention and the Illuminated World," *Brain Pickings*, February 28, 2013, http://www .brainpickings.org/2013/02/28/the-age-of-edison/. See also Ernest Freeberg, *The Age of Edison: Electric Light and the Invention of Modern America* (New York: Penguin, 2013).

range of human hearing: Sound researcher R. Murray Schafer discovered that when American and Canadian students were asked, during meditation in a deeply relaxed state, to sing whatever tone seemed to arise most naturally from the center of their beings, the most frequent response was B natural. Students in Germany and other European countries tended to hum G sharp. In America and Canada, our electricity operates on an alternating current of 60 cycles per second. This resonant frequency corresponds to the B natural tone on the musical scale. In Europe, the electrical current is 50 cycles per second, corresponding musically to G sharp. Exposed for a lifetime to the silent noise in our walls, light fixtures, and appliances, we begin to hum right along with our electricity. R. Murray Schafer, *The Tuning of the World* (Toronto: McClelland and Stewart, 1977), 99.

subdividing an electric current: William J. Broad, "Subtle Analogies Found at the Core of Edison's Genius," *New York Times*, March 12, 1985, http://www .osti.gov/includes/opennet/includes/Understanding%20the%20Atom /World%20Within%20Worlds%20The%20Story%20of%20Nuclear%20 Energy%20Vol.1.pdf.

all pathways are equal: Curiously, this quality of electricity still confounds regular people trying to regulate and legislate the grid. It is very hard for us to wrap our heads around the fact that electricity will take all paths available to it simultaneously, with no preference for what appears logical to us: the shorter, more direct route. As we shall see in chapter 5, this became a real problem during the 2003 blackout.

turned off your TV: Steve Wirt, "The Series Circuit," *Oswego City School District Regents Exam Prep Center*, 1998, http://www.regentsprep.org/Regents /physics/phys03/bsercir/default.htm.

other paths remain open: As resistance in a parallel circuit system increases— when, for example, you plug five power strips, each containing five power cords, into one power strip and plug that into a single wall outlet—current actually increases. This can lead to unexpected conflagrations! This is why, when you were eight and attending mandatory fire safety classes, they urged you not to plug too much stuff into a single outlet.

flickered to light in 1882: Technically, his first actual grid was the one he built to light his Menlo Park laboratory, but Pearl Street was the first public installation.

contemporary 15-watt bulb: If you have seen a vintage "Edison" incandescent in a store or hanging in your favorite bar, you know the relative brightness of these bulbs fairly well. They were dim enough that building codes demanding lightwells were not changed until well into the 1940s, when fluorescent bulbs became more readily available. Carol Willis, *Form Follows Finance: Skyscrapers and Skylines in New York and Chicago* (Princeton, NJ: Princeton Architectural Press, 1995).

populated as lower Manhattan: Pearl Street was designed to light a tiny tranche of lower Manhattan—principally Wall Street but also the offices of the *New York Times*, those of Edison's primary investor, J. Pierpont Morgan, as well as those of his lawyer and principle advocate.

area around Pearl Street: New York had near to 200,000 horses in the late 1800s, each of which laid down a thick layer (up to 30 pounds per day) of horse shit over every cobble of every street, and the valleys between the buildings were filled with sparrows feasting on the remnant grass seeds in this equine effluvia. The flies were a less poetic accompaniment.

other foreign countries: "The Brush Electric Light" *Scientific American* 44 (14). April 2, 1881.

into the 1920s: Hughes (1983).

Edison invented his own: Michael B. Schiffer, *Power Struggles: Scientific Authority and the Creation of Practical Electricity Before Edison* (Cambridge, MA: The MIT Press, 2008), 289.

carbonized bamboo filaments: Maggie Koerth-Baker, *Before the Lights Go Out: Conquering the Energy Crisis Before It Conquers Us* (Hoboken, NJ: John Wiley & Sons, 2012), 18. The electric lightbulb was "invented" by at least twenty-two people before an improved version was successfully commercialized by Thomas Edison in 1879. Henley (2009).

early-autumn rains: Curiously, this is an East Coast and Midwestern problem, where streams have high volumes of water but run over minimal elevation. In the Sierra, where water-powered electricity was very common by the time Appleton got its grid, water runs at low volume but over large shifts in elevation. This, coupled with the gradual melting of snowpack over the spring and summer months, yields a surprisingly even flow of water and thus constant voltage for the lighting systems. Williams (1997), 171.

Wisconsin's fledgling utility: Freeberg (2013), 155. By 1910 the cost of a bulb had come down to 17 cents.

unit of electrical tension: This unit of electrical tension was named after the Italian polymath Alessandro Volta (1734–1827), who invented the first electric battery (the eponymous Voltaic pile) and in the process proved that electricity could be chemically generated. He attracted the attention of Napoleon Bonaparte, who summoned him to Paris in 1801 to present his work at a séance at the Institute of France. By the time Bonaparte was emperor nine years later, he was so enamored with Volta that he refused his attempts to retire from his professorship at the University of Pavia, stating: "a good general ought to die on the field of honor." Beset with honors and flattering invitations from all over Europe, the quiet Volta finally got his wish, retiring in Italy in 1819. John Munro, "Alessandro Volta," in *Pioneers of Electricity; or, Short Lives of the Great Electricians* (London: William Clowes and Sons, Ltd., 1890), 89–102.

pulley system in the place: Eric J. Lerner, "What's Wrong with the Electric Grid?" *Industrial Physicist* 9, no. 5 (November 2003).

electric clocks to slow: On Long Island, it's been a consistent source of jokes that everything is slower there than in the rest of New York State, not because of the leisurely pace of life, but because of the poor quality of their electric power—a New York City second takes something more like 1.005 seconds on Long Island.

sort-of-colorless wash: Freeberg (2013).

either 1,200 or 2,000 volts: Today's American household appliances like toasters and curling irons run on 110 volts, while anything that charges from a USB port is getting 5 volts, and a modern long-distance power line can run easily at 765 kilovolts (kV) (765,000 volts).

Los Angeles, San Jose: David E. Nye, *Electrifying America: Social Meanings of a New Technology, 1880–1940* (Cambridge, MA: MIT Press, 1992), 3.

United States privately owned: The reason that we have private ownership of hot water and of air-conditioning in the United States is because of the very gradual way the two products entered the market. The Soviet Union modernized all at once, and modern laboratory buildings were built on university campuses with a central plant already in place. Time span of development (gradual versus fast) rather than political ideology or a structural bias toward centralization is, in these two cases, the main factor in these systems having been designed differently in different locations.

over municipal sales: If one looks back to the list from *Scientific American* above, about two thirds of installed arc lighting in the United States in 1881 was private.

up and running: Munson (2005), 44.

"almost alone as a central station": Forrest McDonald, *Insull: The Rise and Fall of a Billionaire Utility Tycoon* (Washington, D.C.: Beard Books, 2004), 30.

and electronic trading systems 24/7: Antina Von Schnitzler, "Traveling Technologies: Infrastructure, Ethical Regimes, and the Materiality of Politics in South Africa," *Cultural Anthropology* 28, no. 4 (2013): 670–93.

any inroads at all: Ronald Tobey, *Technology as Freedom: The New Deal and the Electrical Modernization of the American Home* (Berkeley and Los Angeles: University of California Press, 1997).

to make a profit from electricity: Tobey (1997).

power necessary to make it run: Donald MacKenzie and Judy Wajcman, eds., "The Social Shaping of Technology: How the Refrigerator Got Its Hum" (Milton Keynes: Open University Press, 1987).

technologies of any kind: Fred E. H. Schroeder, "More 'Small Things Forgotten': Domestic Electrical Plugs and Receptacles, 1881–1931," *Technology and Culture* 27, no. 3 (1986): 525–43.

flowing in a river: To return for a moment to the atomic physics of it, the thing that an electron desires in an atom is communicated by means of a charge. An electron maintains a minute negative charge and will move toward anything that has a positive charge, as this is what betrays the presence of a slot wherein it can nestle. A whole atom is neutral; it has no charge, whereas an atom after having been broken by an electrical generator is positively charged.

a symptom of it: Today in North America almost all the electricity on the grid is polyphase, and it is kept at an even 60 hertz (Hz) or 60 cycles (reversals) per second. In Europe it's 50 Hz. Japan has both. The number of cycles per second is almost arbitrary. Too low a rate of alternation and incandescent bulbs flicker; you can literally see the packets of energy powering, stopping, and then powering your bulb again as it glows bright, stops, glows bright, stops. With too high a rate of oscillation, motors have a hard time working very well; for the motors' sake, most grids in industrialized countries run at the lowest rate of alternation practicable, given our lighting needs. Most industrializing countries, however, to the degree that they have a national electrical infrastructure at all, pick their preferred rate of oscillation according to the norm prevailing where they import most of their electronics from (usually this is the country that originally colonized them), because the companies making the lightbulbs, toasters, and washing machines for particular markets build them with the appropriate electrical capacities; they are made to run at either 50 or 60 Hz (oscillations per second) and either 110 or 220 volts (intensity of ardor).

using alternating current: "Tesla Life and Legacy—War of the Currents," PBS, http://www.pbs.org/tesla/ll/ll_warcur.html.

competing streetcar lines: According to Munson (2005, p. 32), in 1887 Edison had sold 121 DC central stations, and George Westinghouse, in his first year of business and Edison's main competitor, was working on 68. A year later, in 1888, Edison had installed a total of 44,000 new bulbs, while Westinghouse had installed more than that number (48,000) in October 1888 alone. By 1889, a mere decade after Edison's first viewing of electric bulbs strung in parallel at Menlo Park, Westinghouse had generators running more than 350,000 AC-powered bulbs.

125 cycles per second: Hughes (1983), 128.

machines that used it: Munson (2005), 43, and Schroeder (1986), 530–31.

"polyphase and then the reverse": Hughes (1983), 122. One remarkable thing about Hughes's account of the early processes of electrification is his care in showing the effects that things, rather than people, have on systems design and infrastructural trajectories, including things like business structures, previous investments, little machines, and materials.

or by manila rope: Harold I. Sharlin, "The First Niagara Falls Power Project," *Business History Review* 35, no. 1 (1961): 59–74.

center for the Northeast?: The seeping toxic pit called Love Canal, was one of the worst environmental disasters in American history. It was the result of an early misstep in the history of electrification. A ditch was dug by one Mr. Love in the 1890s to move power between upper and lower Niagara and, even before it was completed it was rendered unnecessary when electricity was chosen as the newest and best means for transmitting power. In the 1920s this half-finished etch into the local landscape was used as an industrial waste dump. Many of the companies that made the products that would allow us to build America into an industrial power house were initially situated near to Niagara, precisely because of the vast quantities of easy-to-access electricity, and their waste went into Love's canal. In the 1950s the area was redeveloped: the canal was covered over with dirt and some houses were built, as was a school for the kids born and raised there. It worked OK, for a while, but in the mid-1970s a particularly intense rainstorm caused the sludge that was buried beneath—22,000 tones of chemical waste—to begin to leach out and percolate up into the gardens and dug basements of these houses. It stank. And the kids, who can be relied upon to touch anything, found themselves with chemical burns on their hands and faces. More on this in chapter 4. Eckardt C. Beck, "The Love Canal Tragedy," *United States Environmental Protection Agency Journal*, January 1979, http://www2.epa.gov/aboutepa/love-canal-tragedy.

"adoption of alternating current": Sharlin (1961), 72.

its lines was begun: "Harnessing Niagara Falls: The Adams Power Station—The Most Famous of Early Hydroelectric Power Stations," *Edison Tech Center: The Miracle of Electricity and Engineering*, 2013, www.edisontechcenter.org/niagara.htm (5/2015).

soon to follow: "The History of the Aluminium Industry," *Aluminium Leader*, accessed November 27, 2015, http://www.aluminiumleader.com/history/industry_history/.

screwed into light sockets: Schroeder (1986), 530–31.

"on the area I was at": From "Niagara Falls Schoellkopf Power Station Disaster, Thursday, June 7th, 1956: A History," *Niagara Frontier*, accessed November 10, 2015, http://www.niagarafrontier.com/schoellkopf.html.

CHAPTER 3: The Consolidation of Power

"45 percent of manufactured products": Richard F. Hirsh, *Power Loss: The Origins of Deregulation and Restructuring in the American Electric Utility System* (Cambridge, MA: The MIT Press, 1999), 12.

(Warren Buffett is number 39): Michael Klepper and Robert Gunther, *The Wealthy 100: From Benjamin Franklin to Bill Gates, A Ranking of the Richest Americans, Past and Present* (New York: Citadel Press, 1996).

absolute control of a market: Hirsh (1999), 16.

and commercial buildings: Munson (2005), 45.

investors to determine: This use of shell corporations, or holding companies, was Enron's number one tool in attracting investment despite being essentially bankrupt.

the standard of American power. It is said that Edison never made another truly innovative intervention in the burgeoning technical world of the late nineteenth century after the death of his first wife in 1884 (see McDonald, 2004). "In 1884, Thomas Edison's world was changing rapidly. Financing for his electric light system was drying up and he was planning to cut his losses and get out of the business altogether. And then Mary died, leaving Edison with three young children (ages 8 to 12), no real job, and no clue about what to do next. He had to borrow $500 to bury his wife." From "Thomas Edison's First Wife May Have Died of a Morphine Overdose," *Rutgers Today*, November 15, 2011, http://news.rutgers.edu/research-news/thomas-edison%E2%80%99s-first-wife-may-have-died-morphine-overdose/20111115#.VfcmIKI_hUx.

the infrastructure that carries it: Batteries don't change this fact, they just shrink down the size of the "infrastructure" for a drastically delimited period of time. The computer I am typing on needs to be plugged in every twelve hours, rather than every three. The "book" you are likely reading this on needs to be plugged in at least once a month. And even the best e-car out there today needs to be plugged in after it's been driven.

"customers at any given moment": Maury Klein, *The Power Makers: Steam, Electricity, and the Men Who Invented Modern America* (New York: Bloomsbury, 2010), 403.

"but its filth and its huge rats": McDonald (2004), 56.

another five hundred private plants: Klein (2010), 401.

"in the hands of a receiver": "Central-station electric service; its commercial development and economic significance as set forth in the public addresses (1897–1914) of Samuel Insull," (Chicago: Privately Printed, 1915), 128. Available at http://archive.org/stream/centralstationelooinsurich/centralstationelooinsurich_djvu.txt.

users of electric current: Klein (2010), 404.

population of the city: Munson (2005), 46.

"2.5¢ per kilowatt-hour in 1909": Munson (2005), 46.

the hundreds of thousands: By lowering his rates Insull made the mass adoption of appliances with their guaranteed twenty-four-hour load affordable, while through his fervent, some would say propagandistic, promotion of the modern home he made refrigerators, freezers, and home hot water desirable.

grid-provided electricity: Munson (2005), 48.

"interest on investment": Klein (2010), 404.

idling for lack of load: Klein writes: " 'If your maximum [load] is very high,' Insull said, 'and your average consumption very low, heavy interest charges will necessarily follow. The nearer your average to your maximum load the closer you approximate to the most economical conditions of production, and the lower you can afford to sell your current.' Insull did not arrive at these insights quickly or all at once, but his experience kept reinforcing one central theme: The way to reduce costs was by increasing output, which meant going after more customers and finding a mix of usages that spread the load across the clock. And the way to get more customers was to lower rates" (2010), 405.

"capacity of 68.5 kilowatts": Munson (2005), 47. Here is Insull's actual version: "Take, for instance . . . a block in a residence district of Chicago which has 193 apartments in it. We have in that block 189 customers, and the number of lamps per customer is between ten and eleven. The kilowatt-hours used per year are 33,000. If you take the customer separate maxima amounting to 68.5 kilowatts you will find that the load factor is only 5.5 percent. All of you know full well that if your entire plant is only in use 5.5 percent of the time it is only a question of time when you will be in the hands of a receiver, but if you take the maximum at the transformers you will find that the maxima of the various customers comes at different times of the day, that instead of the load factor being 5.5 percent it's 10 percent, representing a maximum of 20 kilowatts" (1915).

American electricity industry: Munson (2005), 53.

sport a similar structure: One can imagine that if a gang's territory was supported by government charter, the negotiations at the edges of this territory would be less bloody. See *The Wire*, season 3.

between three and five: Klein (2010), 402.

about 50 percent efficiency: A similar law, called the Betz limit, sets the maximum achievable efficiency of a wind turbine at 59 percent. Peak efficiency is 70 to 80 percent of maximum efficiency, or 41 to 47 percent of the energy available in the wind being converted into electricity. This is slightly higher than what a heat engine can reliably provide, but the real difference is not the efficiency, but the mechanics involved in converting fuel to electric current. Wind is just much less costly and far less polluting than coal.

without boiling it: Hirsh (1999), 56.

around 34 percent efficiency: "What Is the Efficiency of Different Types of Power Plants?" U.S. Energy Information Administration, accessed September 1, 2014, http://www.cia.gov/tools/faqs/faq.cfm?id=107&t=3.

approaching this goal: A better critique, and one discussed at length in chapter 7, is why convert a fuel (wood, coal, gas, etc.) into electricity, transport this electricity vast distances (which involves line loss), and then use it to heat a home, when the same fuel could be used to produce heat without the intervening infrastructure?

in the late 1880s: The second law of thermodynamics was first put to paper in 1824 by Nicolas Léonard Sadi Carnot in *Reflections on the Motive Power of Fire*. He was a French military engineer and physicist who died of cholera at the age of thirty-six. Because people feared contagion, they buried him with most of his writings; this book is his only surviving work.

business in the late 1950s: Indeed, as Jonathan Koomey has pointed out, forecasting, regardless of how complex the model is, does not appear to ever accurately model future events. Personal conversation, December 12, 2010. See also http://www.koomey.com/.

left well enough alone: "The managers' view of technological progress may have
    been influenced by an associated belief in mechanical perfectibility. Trained
    as engineers, managers had an educational background that stressed solving
    problems. This background served them well as they integrated new tech-
    nologies into power production and distribution systems, but it also promul-
    gated the opinion that 'failure is but an anomaly that can be removed in a
    future that is more completely controlled by engineers.' In other words if a
    problem exists the engineer can fix it using the problem-solving approach
    and succeed in perpetuating progress." Hirsh (1989), 125.

America like a sledgehammer: Daniel Barber argues that there was a sense of
    worry about the imbalance between sites of production (of oil) and sites of
    use as early as the late 1940s in the United States and that midcentury solar
    architecture (craftsman) homes are a result of this. Nevertheless the reality
    of this issue didn't hit home until 1973. Daniel A. Barber, "The Post-Oil
    Architectural Imaginary in the 1950s," public lecture, "Lines and Nodes:
    Media, Infrastructure, and Aesthetics," at New York University's Media,
    Culture, and Communication symposium, 2014.

investor-owned leviathans: "U.S. Electric Utility Industry Statistics (2015–2016
    Annual Directory & Statistical Report)," *American Public Power Association*,
    accessed October 26, 2015, http://www.publicpower.org/files/PDFs/USElectric
    UtilityIndustryStatistics.pdf. American Public Power Association sourced
    this information from Energy Information Administration Forms EIA-861
    and 861S, 2013.

and a single substation: "About Us," City of Rancho Cucamonga, accessed
    September 1, 2014, http://www.cityofrc.us/cityhall/engineering/rcmu/aboutus
    .asp.

wealthy, irate adversary: Any California community was, in 2002, granted the
    legal right to administer their own electricity as an aggregate by Assembly
    Bill 117, "An act to amend Sections 218.3, 366, 394, and 394.25 of, and to
    add Sections 331.1, 366.2, and 381.1 to, the Public Utilities Code, relating to
    public   utilities,"   http://www.leginfo.ca.gov/pub/01-02/bill/asm/ab_0101
    -0150/ab_117_bill_20020924_chaptered.pdf.

"the 3,000,000,000 dollar mark": "Samuel Insull Is Victim of Heart Attack in
    Paris," *Berkeley Daily Gazette*, July 16, 1938.

CHAPTER 4: The Cardigan Path

had begun to fail: So much so, that an executive from Virginia Light and Power
    at one point referred to his utility as a "construction company." Hirsh (1989),
    111.

grew too costly to complete: The inordinate expense of building nuclear power plants, each of which was more like a carefully crafted work of art rather than a cookie-cutter prefab job and thus beset by the last-minute modifications and unanticipated delays that come with doing something complex for the first time, made this former truth of the industry's business model stumble as well.

"trucking transportation systems": "Statement of Former U.S. President Carter at Energy Security Hearing Before U.S. Senate Foreign Relations Committee," *Carter Center*, May 12, 2009, http://www.cartercenter.org /news/editorials_speeches/BostonGlobe-energy-security-hearings.html.

decentralized power options: Carter also did a great deal to strengthen America's reliance on fossil fuels, by encouraging more coal use and the exploitation of domestic oil. His problem was energy security and he was very catholic in his approach to solving it. Williams (1997), 325.

commitment to fundamental change: This process might have been inevitable, for as Buckminster Fuller famously pointed out, "All the technical curves rise in tonnage and volumetric size to reach a giant peak after which miniaturization sets in. After that a more economic art takes over which also goes through the same cycle of doing progressively more with less." *Critical Path* (1981).

the project and opted out: Hirsh (1999), 82.

any given billing cycle: The most standard "promotional rate" structure was to charge less the more electricity was used. For example, in 1973 ConEd (New York City's utility) charged "4.4 cents per kWh for each of the first 50 kWh of use, but only 3.9 cents for the next 60 kWh, 3.4 cents for the subsequent 120 kWh and only 2.8 cents for each unit greater than 240 kWh." Richard F. Hirsh, "The Public Utility Regulatory Policies Act," *Powering the Past: A Look Back*, accessed October 1, 2014, http://americanhistory.si.edu /powering/past/history4.htm. (Just as a point of reference, in 2007 the average American home used about 40 kWh per day, or 1,200 kWh per billing cycle, and paid about half as much for the last 1,000 as they did for the first 50). The figure of 40 kWh per day comes from "Average Daily Electricity Usage," *Pennywise Meanderings*, October 2, 2007, http://pinchthatpenny. savingadvice.com/2007/10/02/average-daily-electricity-usage_30740/.

mix of generations: Many early electric power stations, including the Pearl Street Station, recycled their waste heat, usually for district heating. Steam ducts, rather than venting into the air, exited power plants underground and ran through nearby neighborhoods, providing hot water vapor to home and office radiators as far as they went. As power plants got bigger and moved farther away from urban centers, thanks in part to the long distance transmission AC enabled, this secondary use of excess heat by electricity generators was lost. There simply weren't any neighborhoods around and so excess heat was released into the sky.

By 1962 all that had changed: As factories were slowly convinced during the Insull era to buy their electricity instead of making it for themselves, they, too, began to vent waste heat into the atmosphere rather than reusing it. Technically, the steam heated in a cogenerating factory is first used to make electricity and after used for the industrial processes to which the plant is devoted.

all the way down to nothing: Hirsh (1999), 81. Hirsh goes on to state that in 1908, 60 percent of total generation capacity was non-utility, though only some of this was cogeneration. In 1977 that number was slightly over 3 percent and almost all of it was cogeneration.

of their own industrial processes: "Combined Heat and Power: Frequently Asked Questions," United States Environmental Protection Agency, www.epa.gov /chp/documents/faq.pdf.

power to the local utility: Hirsh (1999), 83.

almost all of it was variable: Williams (1997), 328–40.

"Act for Economists of 1978": Hirsh (1999), 125.

"fuel or next planned facility": As quoted in Hirsh (1999), 125.

PURPA era contracts: Randall Swisher and Kevin Porter, "Renewable Policy Lessons from the U.S.: The Need for Consistent and Stable Policies," in *Renewable Energy Policy and Politics: A Handbook for Decision-Making,* edited by Karl Mallon (New York: Earthscan, 2006), 188. Swisher is a former executive director of the American Wind Energy Association (AWEA), 1989–2009; Porter is a former senior analyst for NREL.

promised 1,178 new kWh: This example is taken from Hirsh (1999), 126.

All in 80 MW chunks: Swisher and Porter (2006), 186.

"world's solar power electricity": Hirsh (2014).

"or a wind turbine in California": Paul Gipe, *Wind Energy Comes of Age* (Hoboken, NJ: John Wiley & Sons, 1995), 31.

a tax break of nearly 50 percent: "An incident at Oak Creek Energy systems captures the frenetic pace of the period. Oak Creek, a wind developer near Tehachapi, was making money so fast in the early 1980s that they misplaced a check for $500,000. Auditors eventually found the check—still nego-tiable—nearly a decade later during bankruptcy proceedings." Gipe (1995), 31.

without really understanding why: Cashman's point is that the problem was with the weight and flexibility of the blades—long, flexible blades are not particu-larly good for harvesting the wind. "All the wind turbines that ever worked, they were built by hippies," he pointed out. "The ones in America were built by hippies who actually had masters degrees, bachelors and masters degrees in aeronautical engineering. It was the middle of the Vietnam War and it became clear to them that the only jobs they could get was with Sikorski

Helicopter or some other defense thing which was going to build stuff for Vietnam and they didn't want to do that." "So they were stuck with an expertise and many of them went just back to the land and then they sat there and watched the wind go by and said: 'I know enough about air: I could build a little thing.' And they began building these little wind turbines in different places in the country. And then they thought: 'Well, maybe I should build some for my friends and we should make a little company.' And so they did that." Interview with Tyrone Cashman, March 2011.

10 percent of its in-state generation: These numbers come from "U.S. Wind Energy State Facts," (2015). See also Wieser (2015). "Electric Power Monthly: Table 1.17A. Net Generation from Wind," U.S. Energy Information Administration, August 28, 2015, www.eia.gov/electricity/monthly/epm_table_grapher.cfm?t=epmt_1_17_a. The average American household uses about 1,000 kWh per month; this puts the present wind energy output for California at 1.3 million households and Texas at 3.6 million households.

but that's not news: Zachary Shahan, "Renewable Energy = 13.4 percent of U.S. Electricity Generation in 2014 (Exclusive)," *CleanTechnica*, March 10, 2015, http://cleantechnica.com/2015/03/10/renewable-energy-13-4-of-us-electricity-generation-in-2014-exclusive/. See also "2014 Renewable Energy Data Book" (National Renewable Energy Laboratory of the U.S. Department of Energy, December 2015), http://www.nrel.gov/docs/fy16osti/64720.pdf

predicted to only grow: Roy L. Hales, "Solar & WIND = 53 percent of New U.S. Electricity Capacity in 2014," *CleanTechnica*, February 3, 2015, http://cleantechnica.com/2015/02/03/solar-wind-53-new-us-electricity-capacity-2014/.

power from renewable sources: Dave Levitan, "DOE: U.S. Could Easily Incorporate 80 Percent Renewables in 2050," June 19, 2012, http://spectrum.ieee.org/energywise/energy/renewables/doe-us-could-easily-incorporate-80-percent-renewables-in-2050.

utilities as natural monopolies: Hirsh (1999), 131, quoting Janice Hamrin's "The Competitive Cost Advantages of Cogeneration" (1987).

turning down their thermostats?: Hirsh (1989).

would be something like "reregulation": The term "utility consensus" comes from Hirsh's *Power Loss* (1999).

compliance to its terms: James L. Sweeney, *The California Electricity Crisis* (Stanford, CA: Hoover Institution Press, 2002).

"two years to change a rule": This comes from a personal conversation with Fred Pickel in Los Angeles in 2010. He continued, "whereas if it was allowed to go under a more commercially responsive approach, like the gas industry, when somebody is a bad actor, the rules change. We're no longer doing that kind of thing."

"virtual shopping malls": Bill Keller, "Enron for Dummies," *New York Times,* January 26, 2002, http://www.nytimes.com/2002/01/26/opinion/enron-for-dummies.html.

debt, greed, and risk: If some of Enron's profits were linked to their energy trading, most of their more straightforwardly illegal activity had to do with the exploitation of complex financial tools (derivatives mostly), money laundering, hiding debt in affiliated corporations, and paying government officials to specially manufacture loopholes for them. This, while also allowing individuals within the company, most notably Andy Fastow (the CFO), to maintain huge conflicts of interest in the management of open and hidden assets. Enron was thus first but not foremost an energy company.

"an energy nightmare": Gov. Davis, State of the State Address, January 8, 2001. Quoted in Sweeney (2002), 278.

CHAPTER 5: Things Fall Apart

during a nuclear disaster: Debbie Van Tassel, "Being a Watchdog of FirstEnergy Corp.," *Nieman Reports,* Summer 2004: "The Energy Beat: Complex and Compelling" (June 14, 2004), http://niemanreports.org/articles/being-a-watchdog-of-firstenergy-corp/.

Three Mile Island in 1979: Van Tassel (2004).

20 percent of America's power: Paul L. Joskow and John E. Parsons, "The Future of Nuclear Power After Fukushima" (MIT Center for Energy and Environmental Policy Research, February 2012).

stories of this kind: See especially Oyster Creek Nuclear Generating Station in New Jersey, Nine Mile Point/Unit 1 in New York, and San Onofre (see note below) in California.

doubt that they are old: Davis-Besse began operating in 1978, Vermont Yankee in 1972; San Onofre reactor 1 came online in 1968, and its reactors 2 and 3 (along with the Diablo Canyon reactors) came online in the mid-1980s, making them among the last nuclear power plants in the country to go into operation.

20 percent of its generating capacity: San Onofre—a complex of three nuclear reactors on the rocky California coast between Los Angeles and San Diego—had its number 2 and 3 reactors shuttered in 2012 because of premature wear on over 3,000 newly installed tubes in the replacement steam generators. A year and a half and many attempts to solve the problem later, California decided to shut the plants for good. (Reactor 1 was decommissioned in 1992; it is the most famous of San Onofre's three because its reactor core was accidentally installed upside down back in 1977 when the station was first built.) Said David Freeman (former head of the California

Power Authority) of San Onofre and Diablo Canyon nuclear power plants: they are "disasters waiting to happen: aging, unreliable reactors sitting near earthquake fault zones on the fragile Pacific Coast, with millions or hundreds of thousands of Californians living nearby" (quoted in Eric Wesoff, "PG&E Study: Diablo Canyon Nuclear Plant Is Earthquake-Safe Despite Newly Detected Faults," *Greentech Media*, September 15, 2014, http://www .greentechmedia.com/articles/read/PGE-Diablo-Canyon-Nuclear-Plant-is-Earthquake-Safe-Despite-Newly-Detected). True enough, but with the closure of San Onofre, Southern California lost 20 percent of its electrical generation capacity. Should Diablo Canyon, the last of California's nuclear reactors, also be closed (it *was* unfortunately built directly on top of a geologic fault line), California will lose an additional 7 percent of its electrical generating capacity—about 2.2 million homes' worth.

modern electric company: "U.S. Electric Utility Industry Statistics (2015–2016 Annual Directory & Statistical Report)," (2015).

than by producing it: "The Changing Structure of the Electric Power Industry 2000: An Update" (Washington, DC: Energy Information Administration, Office of Coal, Nuclear, Electric and Alternate Fuels, U.S. Department of Energy, October 2000).

in good working order: Van Tassel (2004).

more growth than is acceptable: See "Frequently Asked Questions (FAQs): Tree Trimming and Vegetation Management Landowners," *Federal Energy Regulation Commission*, accessed May 1, 2015, http://www.ferc.gov /resources/faqs/tree-veget.asp.

and burst into flames: PG&E was allegedly responsible for at least fifty other fires in California, four of which were serious, before the Trauner Fire case was filed. This is according to Tom Nadeau, *Showdown at the Bouzy Rouge: People v. PG&E* (Grass Valley, CA: Comstock Bonanza Press, 1998), 14. This is an exceptional book that tells the story of the people of Nevada County attempting to hold PG&E responsible for their inactions. The company has more employees than Nevada County has people, and their budget for legal defenses is forty times that of Nevada County. But the county won.

national electric infrastructure: Squirrels form a nasty second in this regard; see chapter 7.

"and we will melt": John P. Coyne, "Boom Signaled Power-Line Arc in Walton Hills," *The Plain Dealer*, August 24, 2003, http://web.archive.org/web /20071217151746/http://www.cleveland.com/blackout/index.ssf?/blackout /more/1061717873203660.html.

very quickly indeed: An electric charge in a vacuum moves at the speed of light; domesticated electricity on earth moves somewhat slower due to the resistance in the wires. Nevertheless, it still moves very fast.

a fork in an outlet: Humans are relatively good conductors, which is why we have built an electrical system that separates us from it by all sorts of poor conductive materials, like the air, rubber, glass, or ceramic. The reason young Adam Muha was warned with such vehemence out of his driveway by a lineman, who knew more about the ways of electricity than the teen, was because when electricity is arcing it will seek to land on a good conductor that's also ideally a bit pointy, like a tree, or a tall boy, or, in the American West, a cow (hundreds of which are killed by lightning every year).

"not anticipated or controlled": Lerner (2003), 8.

within which this machine functions: This is true for airplanes as well; there was simply no way to take into account, before 2001, the fact that some people might decide to use airplanes as bombs. This choice by Al Qaeda falls into a different model of organizational disaster: the Black Swan, in which unimaginable things do exist and very often these are the selfsame things that surprise us and destroy the systems we have put into place. The global cultural environment was not initially included in the Swiss cheese model for airline safety. Now it is. And we stand in long security lines and are subject to two entirely new government agencies (TSA and Homeland Security) as a result.

thoroughly man-made: It does seem that in all of this, there was a single criminal act. Somebody, under cover of darkness, stole the top of that very first tree. So when the government task force field team investigating the blackout was sent round to the Muhas' place, they found the offending tree nicely sawed up and stacked in piles—a task performed by FirstEnergy's crews as part of the postblackout cleanup. But when these logs were unstacked and measured, they amounted to only 42 feet of tree—the very top was gone! While not exactly accusing anybody of anything, the investigating team noted in the "Final Report on the August 14, 2003 Blackout in the United States and Canada: Causes and Recommendations" (U.S.-Canada Power System Outage Task Force, April 2004), p. 60, that "portions of the tree had been removed from the site," a fact that made it difficult to "determine the exact height of the line contact, the measured height is a minimum and the actual contact was likely three to four feet higher."

In other words, this missing arboreal crown served a purpose. Without it, it was impossible to prove criminal negligence in FirstEnergy's admittedly poky approach to tree pruning. The evidence necessary to make such a case had been tampered with. Admitting to nothing, FirstEnergy spokesman Ralph DiNicola retorted: "If something was missing we can surmise someone grabbed some firewood." And perhaps he is right; somebody might have just carted off that last, critical four-plus feet of treetop to burn in their woodstove. But why not take the rest of the tree? After all, there were another 42 feet of nicely hewn trunk ready to be stacked, to age, and to become heat of their own, one long winter night. Though eyebrows were raised by this seemingly innocuous theft of evidence from the origin point of the blackout, it is almost more remarkable for being the only moment of maleficence in the entire debacle. See Peter Krouse, Teresa Dixon Murray, and John Funk, "Top of tree in blackout investigation is missing remnants found under FirstEnergy line," *Plain Dealer*, November 22, 2003: C1.

afternoon the East went dark: It wasn't the weather that brought down the grid; it wasn't the wind, and it wasn't even raining. And though it was hot, in the high 80s across most of the blacked-out region, none of the first three lines were overloaded—they were not being asked to carry more electricity than they could bear. Air conditioners may have been sucking up electrons and churning out cool air everywhere in the East, and this demand for electricity certainly meant the system was operating under stress, but they were not deemed causal.

"use their own tunes": Lerner (2003), 10.

40 tons of chlorine bleach for free: Joe Taffe, personal conversation, May 10, 2011.

like pork bellies or pig iron: From the point of view of the consumer, electrons still bear little resemblance to pork bellies and pig iron. This is one of the problems discussed in chapter 6.

homes in Columbus, Ohio: This is using the crude measure of twenty-five 100-watt lightbulbs per household. The example is also from Lerner (2003).

25 percent of all energy trading: "Electricity 101: Frequently Asked Questions," *Office of Electricity Delivery & Energy Reliability*, accessed September 23, 2015,http://energy.gov/oe/information-center/educational-resources/electricity-101#ppl1.

care or recipient of cash: Hertzog (2013).

buoyed up and smoothed out: For another version of this same story, one that blames the blackout almost entirely on a lack of reactive power (VARS) on the grid, see Jane Bennett, *Vibrant Matter: A Political Ecology of Things* (Durham, NC: Duke University Press, 2010): 24–28.

in the summer of 1996: Robert Peltier, "How to Make VARs—and a Buck," *PowerMagazine*,June 15,2007,http://www.powermag.com/how-to-make-varsand-a-buck/?pagenum=1.

CHAPTER 6: Two Birds, One Stone

station after the event: Charlie Wells, "Houston Woman Thelma Taormina Pulls Gun on Electric Company Worker for Trying to Install 'Smart Meter,'" *New York Daily News,* July 19, 2012, http://www.nydailynews.com/news/national

/houston-woman-thelma-taormina-pulls-gun-electric-company-worker-install-smart-meter-article-1.1118051.

which were watching Shrek 2: "Researchers Claim Smart Meters Can Reveal TV Viewing Habits," Metering.com, September 21, 2011, http://www.metering .com/researchers-claim-smart-meters-can-reveal-tv-viewing-habits/. For the research conducted at the University of Washington, see Antonio Regalado, "Rage Against the Smart Meter," *MIT Technology Review*, April 26, 2012, http://www.technologyreview.com/news/427497/rage-against-the-smart-meter/. And for readers of German: Prof. Dr.-Ing U. Greveler, Dr. B. Justus, and D. Löhr, "Hintergrund und Experimentelle Ergebnisse Zum Thema 'Smart Meter und Datenschutz'" (Fachhochschule Münster University of Applied Sciences, September 20, 2011), https://web.archive.org/web /20121117073419/http://www.its.fh-muenster.de/greveler/pubs/smartmeter _sep11_v06.pdf.

other in any substantial way: On the utility side the claims sounded like this: "The accuracy of Smart Meters, both in development and practice, has been confirmed to improve on the older electro-mechanical meter technology. All meters, regardless of technology and design, are required to meet national standards such as ANSI C12 for meter accuracy and operation before being installed." "Smart Meters and Smart Meter Systems: A Metering Industry Perspective" (Edison Electric Institute, Association of Edison Illuminating Companies, Utilities Telecom Council, March 2011), http://www.eei.org /issuesandpolicy/grid-enhancements/documents/smartmeters.pdf.

   Nevertheless, a great many people noticed that their bills after the new meters were higher. Tom Zeller Jr., "'Smart' Electric Meters Draw Complaints of Inaccuracy," *New York Times*, November 12, 2010, http:// www.nytimes.com/2010/11/13/business/13meter.html. See also Katherine Tweed, "Are Traditional Electricity Meters Accurate?" *Greentech Media*, March 30, 2010, http://www.greentechmedia.com/articles/read/are-traditional-elecricity-meters-accurate.

will rise again anytime soon: "Energy and Technology: Let There Be Light," *Economist*, January 17, 2015, 11.

meters as of 2014: "Frequently Asked Questions: How Many Smart Meters Are Installed in the United States, and Who Has Them?" U.S. Energy Information Administration, accessed November 1, 2014, http://www.eia .gov/tools/faqs/faq.cfm?id=108&t=3.

electricity use at the same time: Martin LaMonica, "GreenBiz 101: What Do You Need to Know About Demand Response?" *GreenBiz,* April 29, 2014, http:// www.greenbiz.com/blog/2014/04/29/greenbiz-101-what-do-you-need-know-about-demand-response.

to right about 2 percent: "A typical coal-fired electrical plant might be 38 percent efficient, so a little more than one-third of the chemical energy content of the fuel is ultimately converted to usable electricity. In other words, as much as 62 percent of the original energy fails to find its way to the electrical grid. Once electricity leaves the plant, further losses occur during delivery. Finally, it reaches an incandescent lightbulb where it heats a thin wire filament until the metal glows, wasting still more energy as heat. The resulting light contains only about 2 percent of the energy content of the coal used to produce it. Swap that bulb for a compact fluorescent and the efficiency rises to around 5 percent—better, but still a small fraction of the original." This comes from "What You Need to Know About Energy: Sources and Uses," *The National Academies Press*, http://www.nap .edu/reports/energy/sources.html. Also, though his math is wrong, Rob Rhinehart has a delightful article on the same subject. "How I Gave Up Alternating Current," *Mostly Harmless*, August 3, 2015, http://robrhinehart .com/?p=1331&utm_source=Daily+Lab+email+list&utm_campaign= 445526c8cf-dailylabemail3&utm_medium=email&utm_term=0_ d68264fd5e-445526e8ef-364971681.

to deal with every day: "In spite of the increasing number of pubs with large-screen televisions, 71 per cent of football fans would watch a final involving England at home or at a friend's house, a survey . . . found. This would lead to a massive surge in electricity during half time intervals and after the final whistle is blown, as people head to the fridge for a beer or the kettle for a cup of tea. The phenomenon, known as a TV pickup, occurs most days during popular programmes, but big football matches result in a heavier demand. The expected increase in electricity usage for England v USA will be 1,200 megawatts at half time and around 1,100 megawatts at the final whistle. The predicted surge if England gets to the final and the game goes to penalties would beat the previous record for a TV programme, set after England lost on spot kicks to West Germany in the 1990 World Cup when a 2,800 MW demand was imposed by the ending of the penalty shootout in the England v West Germany FIFA World Cup semifinal." "National Grid Anticipates Power Surges during World Cup," *Telegraph*, June 11, 2010, http://www.telegraph.co.uk/news/earth/ energy/7819443/National-Grid-anticipates-power-surges-during-World-Cup.html. See also Mark Raby, "Tea Time in Britain Causes Predictable, Massive Surge in Electricity Demand," *Geek*, January 7, 2013, http://www.geek.com/news/tea-time-in-britain-causes-predictable-massive-surge-in-electricity-demand-1535023/.

a moment of peak revenue: The bad news about peak shaving is that now that utilities have finally found a way to make money off peak demand with smart meters, eliminating peak demand now means taking a hit at their bottom line. Charging for the most expensive power at the most expensive time of day was an economic win for them. Bill McKibben, "Power to the People,"

New Yorker, June 29, 2015, http://www.newyorker.com/magazine/2015/06 /29/power-to-the-people. See also John Farrell, "Utilities Cry 'Fowl' Over Duck Chart And Distributed Solar Power," *CleanTechnica*, July 21, 2014, http://cleantechnica.com/2014/07/21/utilities-cry-fowl-over-duck-chart-and-distributed-solar-powercrying-fowl-or-crying-wolf-open-season-on-the-utilitys-solar-duck-chart/. And Katie Fehrenbacher, "This Startup Just Scored a Deal to Install a Massive Number of Tesla Grid Batteries," *Fortune*, June 4, 2015, http://fortune.com/2015/06/04/advanced-microgrid-solutions/.

"Effect Hits Santa Cruz": Gary L. Hunt, "The Bakersfield Effect Hits Santa Cruz," *Tech & Creative Labs*, August 29, 2010, http://www.tclabz .com/2010/08/29/the-bakersfield-effect-hits-santa-cruz/.

like a cash grab: Jack Danahy, "Smart Grid Fallout: Lessons to Learn from PG&E's Smart Meter Lawsuit," *Smart Grid News*, November 13, 2009, http://www.smartgridnews.com/story/smart-grid-fallout-lessons-learn-pge-s-smart-meter-lawsuit/2009-11-13, for individual customer complaints see: https://sites.google.com/site/nocelltowerinourneighborhood/home/wireless-smart-meter-concerns/smart-meter-consumers-anger-grows-over-higher-utility-bills.

digital smart meters: Jesse Wray-McCann, "Householders Shielding Homes from Smart Meter Radiation," *Herald Sun*, April 9, 2012, http://www. heraldsun.com.au/ipad/householders-shielding-homes-from-smart-meter-radiation/story-fn6bfm6w-1226321653862.

commissioners' residences: Anjeanette Damon, "Smart Meters Spawn Conspiracy Talk: They Know What You're Watching on TV!," *Las Vegas Sun*, March 8, 2012,http://m.lasvegassun.com/news/2012/mar/08/smart-meters-spawn-conspiracy-theories-they-know-w/.

"other ball-shaped organs": "Assessment of Radiofrequency Microwave Radiation Emissions from Smart Meters" (Santa Barbara, CA: Sage Associates, January 1, 2011), http://sagereports.com/smart-meter-rf/?page_id=196.

efficient appliances: Margaret Taylor et al., "An Exploration of Innovation and Energy Efficiency in an Appliance Industry" (Ernest Orlando Lawrence Berkeley National Laboratory, March 29, 2012), http://eetd.lbl.gov/sites/all /files/an_exploration_of_innovation_and_energy_efficiency_in_an_appliance_ industry_lbnl-5689e.pdf.

policies in place: Jocelyn Durkay, "Net Metering: Policy Overview and State Legislative Updates," *National Conference of State Legislatures*, September 26,2014,http://www.ncsl.org/research/energy/net-metering-policy-overview-and-state-legislative-updates.aspx.3.

the initial shattering: Jeffrey Sparshott, "More People Say Goodbye to Their Landlines," *Wall Street Journal*, September 5, 2013, http://www.wsj.com /articles/SB10001424127887323893004579057402031104502.

$21 million: Glenn Fleishman, "Stick a Fork in It: A Broadband over Powerline Post Mortem," *Ars Technica*, October 23, 2008, http://arstechnica.com /uncategorized/2008/10/stick-a-fork-in-it-a-broadband-over-powerline-post-mortem/. "Data signals are blocked from the high- and medium-voltage lines over which BPL works by the transformers that step that voltage down for household or business use. To install smart meters, a bypass has to be put in place that sucks the data from the higher-voltage lines and feeds it out on the wire that heads into the home, using home powerline technology. Some companies tried to do this with WiFi on the pole; others put in relatively inexpensive shunts. But it's extremely labor-intensive; more akin—but far less work—to putting in fiber optic to a home than in overlaying data onto cable or telephone wires."

would not use again: Mark Jaffe, "Xcel's SmartGridCity Plan Fails to Connect with Boulder," *The Denver Post*, October 28, 2012, http://www.denverpost. com/ci_21871552/xcels-smartgridcity-plan-fails-connect-boulder.

"Give me a blinking break": April Nowicki, "Boulder's Smart Grid Leaves Citizens in the Dark," *Greentech Media*, March 18, 2013, http://www. greentechmedia.com/articles/read/Boulders-Smart-Grid-Leaves-Citizens-in-the-Dark.

"Stupid Customer' pilot": Jesse Berst, "SmartGridCity Meltdown: How Bad Is It?" *Smart Grid News*, August 8, 2010, http://www.smartgridnews.com/ story/smartgridcity-meltdown-how-bad-it/2010-08-03.

positions on the matter: Randy Houson, business technology executive for Xcel Energy, public speech,Washington, D.C., September 22, 2009.

"very hot outside": Stephen Fairfax wrote this online comment in response to Jesse Berst's "SmartGridCity Meltdown: How Bad Is It?" (2010).

never signed up for: Apparently, the day before the public announcement of the selection of Boulder as the site of the SmartGridCity project, city council members had no idea the project was being undertaken. There had been some serious discussion about municipalizing the utility before their being chosen to get a smart grid, and at least one Boulder resident supposes that part of the reason they were picked was to keep them from defecting. In the end, though, municipalizing the grid is precisely what the city did do. In the words of Steve VanderMeer: "The city of Boulder has been unhappy with Xcel for years and has been contemplating municipalizing their system during much of that time ... By coincidence, I had lunch with a high Boulder official the week of the SmartGridCity announcement. Turns out the city of Boulder had no idea this was coming until the day before the announcement! There was no prior planning, no collaboration. When I asked this official to speculate on the motives of Xcel's announcement, this person wondered out loud whether SmartGridCity was a bone that Xcel was throwing at Boulder in an effort to dissuade further discussions about parting ways with them. Based on current discussions within the city to not renew the franchise agreement, it appears that if this was a motive of Xcel's, it didn't work." Berst (2010).

customers with the undertaking: Mark Jaffe, "PUC Reduces Amount Xcel Can Charge for SmartGrid Project," *Denver Post,* January 5, 2011.

permanent and total: Lewis Milford, "The End of the Electric Utilities? The Industry Thinks So Too," *The Huffington Post,* September 25, 2013, http://www.huffingtonpost.com/lewis-milford/electric-utilities-future_b_3660311.html.

bill us on the other side: Michael Kanellos, "Another Way to Look at the Utility Death Spiral," *Forbes,* September 29, 2014, http://www.forbes.com/sites/michaelkanellos/2014/09/29/another-way-to-look-at-the-utility-death-spiral/. Nicholas Brown, "Will Renewable Energy Cause a Utility 'Death Spiral'? No Need for That," *CleanTechnica,* June 24, 2014, http://cleantechnica.com/2014/06/24/will-renewable-energy-cause-utility-death-spiral/. Martin LaMonica, "Efficiency Group Says 'Utility Death Spiral' Talk Is Overblown," *Greentech Media,* June 13, 2014, https://www.greentechmedia.com/articles/read/utility-death-spiral-talk-overblown-says-efficiency-group. Zachary Shahan, "Warren Buffett: Utility Death Spiral Is Bull S°&^," *CleanTechnica,* March 25, 2014, http://cleantechnica.com/2014/03/25/warren-buffett-utility-death-spiral-bs/.

their power from: The Santa Clara County Jail ("the green prison") and the University of California, San Diego campus are two examples of this sort of islandable institutional microgrid.

via the Internet: "Vision vs. Reality," *Denver Post,* accessed December 1, 2014, http://www.denverpost.com/portlet/article/html/imageDisplay.jsp?contentItemRelationshipId=4738191.

all work together: Berst (2010).

"energy, or coal-fired": This is impossible since biofuels for airplanes, while in development especially in Brazil and while theoretically sustainable, are barely renewable. Airplanes never run on coal.

largest bankruptcy in U.S. history: Paul Starr, "The Great Telecom Implosion," *American Prospect,* September 8, 2002, https://www.princeton.edu/~starr/articles/articles02/Starr-TelecomImplosion-9-02.htm.

"environmental savings": Fleishman (2008).

the polar vortex: Andrea Thompson and Climate Central, "2015 May Just Be Hottest Year on Record," *Scientific American,* August 20, 2015, http://www.scientificamerican.com/article/2015-may-just-be-hottest-year-on-record/.

soothe their woes: "In 1970, two thirds of new home owners kept cool without central air-conditioning: today, central air-conditioning is a standard feature in 90 percent of new homes, even in temperate climates," Janda (2011), 18.

"grids will likely prosper": Fleishman (2008).

"highways to free up bottlenecks": Massoud Amin, "System-of-Systems Approach," in *Intelligent Monitory, Control, and Security of Critical Infrastructure Systems*, edited by Elias Kyriakides and Marios Polycarpou (New York: Springer, 2015), 337.

"is currently [2008] at 10 to 15 percent.": Amin (2015), 337.

"demand is highest": Chris King, "How Smart Meters Help Fight Power Outages," *Gigaom*, July 5, 2012, https://gigaom.com/2012/07/05/how-smart-meters-help-fight-power-outages/. See also "Energy Wise Rewards: Frequently Asked Questions," *Pepco*, accessed September 24, 2015, https://energywiserewards.pepco.com/md/faq/index.php.

power companies started to fail: "Memorandum: General Information on Utility Bankruptcy" (Montana Public Service Commission, July 9, 2003), http://psc.mt.gov/consumers/energy/pdf/BroganUtilityBankruptcy.pdf.

maybe even some lines: Morgan Stanley and Citigroup are major holders of electricity assets at present, including generation. Lerner (2003), 12.

including Dallas–Fort Worth: "Explaining Oncor Electricity," *Bounce Energy*, accessed November 13, 2015, https://www.bounceenergy.com/articles/texas-electricity/oncor-electricity.

CHAPTER 7: A Tale of Two Storms

works in this way: "The Energy Cloud," *Navigant Research*, June 3, 2014, www.navigantresearch.com/webinar/the-energy-cloud-1.

took to acquire it: NATO dropped special "blackout" nets on the city during the bombing; these fall from planes and catch on electric wires, shorting them all out and ideally propagating a blackout from the point of contact through the grid as a whole.

fully coupled to the jet stream: Jeff Halverson, "Superstorm Sandy: Unraveling the Mystery of a Meteorological Oddity," *Washington Post*, October 29, 2013, https://www.washingtonpost.com/blogs/capital-weather-gang/wp/2013/10/29/superstorm-sandy-unraveling-the-mystery-of-a-meteorological-oddity/.

miles in the nation: The area called the Northeast megalopolis stretches across less than 2 percent of the U.S. land mass but contains 17 percent of the country's population. John Rennie Short, *Liquid City: Megalopolis and the Contemporary Northeast* (New York: Routledge, 2007), 23.

stranded, they were immobilized: "Two-thirds of those living in the affected region reported that they lost power. Forty percent of those who lost power report outages of a week or more. The storm left 44 percent in the hardest hit regions without heat and nearly half those, 49 percent, were without heat for more than a week. Thirteen percent in these areas lost water with 36 percent having lost water for more than a week . . . half of affected Americans say they had trouble accessing fuel." Trevor Tompson et al., "Resilience in the Wake of Superstorm Sandy" (Associated Press–NORC Center for Public Affairs Research, June 2013), 3.

when the grid breaks: "Economic Benefits of Increasing Grid Resilience to Weather Outages" (2013).

military and police forces: Ted Koppel, *Lights Out: A Cyberattack, A Nation Unprepared, Surviving the Aftermath* (New York: Crown Publishing, 2015).

"imposed on ourselves": Lovins and Lovins (1982), 1.

"our whole way of life": Lovins and Lovins (1982), 1.

a single cable failed: Thomas Gaist, "The Detroit Blackout," *World Socialist*, December 4, 2014, http://www.wsws.org/en/articles/2014/12/04/pers-d04.html.

"diversity over homogeneity": Lovins and Lovins (1982), 184.

reports, and pontificating pundits: Many academics and even some bloggers are turning against resilience as an approach to maintaining systems (including individual human psychology), as it presupposes a world in crisis. To be resilient you have to accept that things are going to be broken in the first place, so the question becomes what sort of life we have constructed for ourselves that we can't manage the basics of security and have to just build and plan with the assumption of inevitable catastrophe undergirding everything we do. See especially Sandy Zelmer and Lance Gunderson, "Why Resilience May Not Always Be a Good Thing: Lessons in Ecosystem Restoration from Glen Canyon and the Everglades," *Nebraska Law Review* 87, no. 4: 2008.

"Weather Outages": "Economic Benefits of Increasing Grid Resilience to Weather Outages" (2013).

"high- and low-tech solutions": "Economic Benefits of Increasing Grid Resilience to Weather Outages" (2013), pages 17, 12, and 5, respectively.

by the surrounding area: "In the Aftermath of Superstorm Sandy: Message from President Stanley," *Stony Brook University*, accessed December 1, 2014, http://www.stonybrook.edu/sb/sandy/.

devastated Japan in 2011: Asmus (2012).

pumps, and power on: The quote is from Peter Asmus, Alexander Lauderbaugh, and Mackinnon Lawrence, "Executive Summary: Market Data: Microgrids" (*Navigant Research*, 2013), http://www.navigantresearch.com/wp-assets/uploads/2013/03/MD-MICRO-13-Executive-Summary.pdf, 2. See also Silvio Marcacci, "Over 400 Microgrid Projects Underway En Route to $40 Billion Market," *CleanTechnica*, April 3, 2013, http://cleantechnica.com/2013/04/03/over-400-microgrid-projects-underway-en-route-to-40-billion-market/.

air conditioners or clothes dryers: "About Microgrids," Microgrids at Berkeley
 Lab, U.S. Department of Energy, accessed September 28, 2015, https://
 building-microgrid.lbl.gov//about-microgrids.
$8 million a year in electricity bills: "Energy and Technology: Let There Be
 Light" (2015), 11.
"strain on military budgets": Pew Trusts, "Military Clean Energy Innovation:
 Pew," 2011, https://www.youtube.com/watch?v=HiOvfdYsQEE."
"but more importantly in blood.": Peter Byck, Carbon Nation, 2010.
eight gallons a day to four: Jenkins (2011).
batteries needed to run them: Roy H. Adams III, Martin F. Lindsey, and Anthony
 Marro, "Battlefield Renewable Energy: A Key Joint Force Enabler," Joint
 Force Quarterly, no. 57, 2nd Quarter: Stability and Security Operations
 (2010), 45. These devices include "computer displays, infrared sights, global
 positioning systems, night vision, and a variety of other sensor technologies."
 Theodore Motyka, "Hydrogen Storage Solutions in Support of DoD
 Warfighter Portable Power Applications," WSTIAC Quarterly 9, no. 1
 (2009), 83.
third of this weight: Phillip Jenkins, "Lightweight, Flexible Photovoltaics for
 Mobile Solar Power," Ninth International Energy Conversion Conference,
 San Diego, CA, July 31–August 3, 2011.
disposable batteries: Motyka (2009), 83.
"a cost of $700,000": Adams III et al. (2010), 45. These batteries are called
 "primary" because they're not rechargeable, but are thrown away. This
 article is from 2010, and it's citing an article from 2009, which is citing a
 report from 2007, so it's possible that a brigade is no longer throwing out 7
 tons of batteries for a 72-hour mission.
"wrong with this picture": Carbon Nation (2010).
"relative to sustained outages": LaCommare and Eto (2005), as quoted in
 "Economic Benefits of Increasing Grid Resilience to Weather Outages"
 (August 2013), 18.8.
electricity to work: "More than ever our operating forces rely on the use [of]
 electric power to support critical command and control functions; intelligence, surveillance, and reconnaissance assets." Adams III et al. (2010),
 43–44.
added twenty-one microgrids: "The DoD's interest in improving energy security
 through microgrid technology stems from its heavy reliance upon all forms
 of fossil fuels," says Peter Asmus, principal research analyst with Navigant
 Research. "In addition, the DoD has reexamined the existing electricity
 service delivery model in the United States, and has concluded that the best
 way to bolster its ability to secure power may well be through microgrid
 technology it can often own and control." "Microgrids for Military Bases to
 Surpass $377 Million in Annual Market Value by 2018," Navigant Research,

May 10, 2013, http://www.navigantresearch.com/newsroom/microgrids-for-military-bases-to-surpass-377-million-in-annual-market-value-by-2018.

already operate in the United States: "New Jersey Becomes Latest State to Invest in Microgrids," *GreenBiz*, (September 6, 2013), http://www.greenbiz.com /blog/2013/09/06/new-jersey-becomes-latest-state-invest-microgrids. According to a 2013 report from Navigant Research, there are now more than fifty U.S. military bases that are operating, planning, or testing microgrids. See also Koch (2013).

to 578 megawatts: Marcacci (2013).

ventilation by 40 to 75 percent: William M. Solis, "Defense Management: DoD Needs to Increase Attention on Fuel Demand Management at Forward-Deployed Locations" (United States Government Accountability Office, February 2009), http://www.gao.gov/new.items/d09300.pdf. "By foaming tents, we go from a 50-ton cooling unit down to an 8-ton cooling unit, which reduces our power consumption tremendously. We figured that within approximately 10 months we will pay for everything. So we are actually saving soldiers' lives, giving them a more comfortable place to stay and saving money and saving the environment all in one fell swoop."

"the security environment": Adams III et al. (2010), 43–49.

too hot, or rigid: Jenkins (2011). This is similar to the problem of black smoke produced by coal burners on ships before World War I, which allowed enemies to spot them before they crossed the horizon. Technology has gotten better across the board and the fuels in use need to adapt to this. This links to the desire for a solar panel that isn't square, isn't shiny, and isn't black. There is a desire for camouflage, for invisibility that extends now into the IR range. Heat signatures are a sort of visibility given our new technological vision.

"system's control unit": Adams III et al. (2010), 46.

any piece of it: A couple of the Straws' neighbors to the south, the Yoxulls, have been off the grid since they built their house eighteen years ago because they didn't want to pay the utility $20,000 to extend the existing system of lines to a new pole.

"of a typical tactical operations center": Adams III et al. (2010), 46.

"meet power requirements": According to Adams III et al., "[the] inherent advantage of [renewables for deployed off-grid operations] over conventional petroleum-fueled systems is that combined with demand reduction, they greatly reduce and even eliminate the need to provide fuel logistic to remote sites, saving manpower, funds, and most importantly decreasing the risk to forces delivering supplies over contested lines of communication." They go on: "In addition, solar-PV and wind technologies offer significant inherent security features in that they are quiet and have low thermal signatures" (2010), 44–45.

"such as biomass conversion": THEPS is the name of a system made by the Arlington-based SkyBuilt that more recently has been working with Lockheed Martin on the Integrated Smart-BEAR Power System (BEAR is the name for the base grid), the point being that the names and specifics of these things change, but for ten years at least, the basic principles have remained the same. See this great article: "Frontline Commanders Requesting Renewable Power Options," *Defense Industry Daily*, January 24, 2012, http://www.defenseindustrydaily.com/commanders-in-iraq-urgently-request-renewable-power-options-02548/.

power the cookstoves: A biorefinery the size of a semi can "process the daily waste of 500 soldiers and generate 60 kW." Adams III et al. (2010), 46.

and $850,000 developing: Adams III et al. (2010), 46, and Solis (2009). More recently, see also Franklin H. Holcomb, "Waste-to-Energy Projects at Army Installations" (U.S. Army Corps of Engineers, January 13, 2011), http://energy.gov/sites/prod/files/2014/03/f11/waste_holcomb.pdf, "U.S. Military Waste to Energy & Fuel Gasification Prototype V2.0," December 7, 2012, http://www.waste-management-world.com/articles/2012/12/u-s-military-waste-to-energy-fuel-gasification-prototype-v2-0.html, and Don Kennedy, "Garbage to Fuel: Trash-to-Fuel Generator, Battle-Tested in Iraq, Shows Long-Term Potential," *Army AL&T Magazine*, April–June 2013, 138–141.

transporting other gasoline around: *Carbon Nation* (2010).

extractive technologies: Timothy Mitchell, *Carbon Democracy: Political Power in the Age of Oil* (Brooklyn, NY: Verso Books, 2011).

substation's large transformers: Rebecca Smith, "PG&E Silicon Valley Substation Is Breached Again," *Wall Street Journal*, August 28, 2014, http://www.wsj.com/articles/pg-es-metcalf-substation-target-of-construction-equipment-theft-1409243813.

computerized infrastructure: "Electric Disturbance Events (OE-417)," *Office of Electricity Delivery & Energy Reliability*, accessed September 30, 2015, http://www.oe.netl.doe.gov/oe417.aspx. The count is mine.

in just three months: Jon Mooallem, "Squirrel Power!" *New York Times*, August 31, 2013, http://www.nytimes.com/2013/09/01/opinion/sunday/squirrel-power.html?_r=0. The litany continues: "Squirrels cut power to a regional airport in Virginia, a Veterans Affairs medical center in Tennessee, a university in Montana and a Trader Joe's in South Carolina. Five days after the Trader Joe's went down, another squirrel cut power to 7,200 customers in Rock Hill, S.C., on the opposite end of the state. Rock Hill city officials assured the public that power outages caused by squirrels were 'very rare' and that the grid was 'still a reliable system.' Nine days later, 3,800 more South Carolinians lost power after a squirrel blew up a circuit breaker in the town of Summerville."

"In Portland, Ore., squirrels got 9,200 customers on July 1; 3,140 customers on July 23; and 7,400 customers on July 26. ('I sound like a broken record,' a spokesman for the utility said, briefing the press for the third time.) In Kentucky, more than ten thousand people lost power in two separate P.O.C.B.S. [power outages caused by squirrels] a few days apart. The town of Lynchburg, Virginia, suffered large-scale P.O.C.B.S. on two consecutive Thursdays in June. Downtown went dark. At Lynchburg's Academy of Fine Arts, patrons were left to wave their lighted iPhone screens at the art on the walls, like torch-carrying Victorian explorers groping through a tomb."

"One June 9, a squirrel blacked out 2,000 customers in Kalamazoo, Mich., then 921 customers outside Kalamazoo a week later. A local politician visited the blown transformer with her children to take a look at the culprit; another witness told a reporter, 'There was no fur left on it. It looked like something from C.S.I.' She posted a photo of the incinerated animal to her Facebook page."

"causing frequent power outages": Mooallem (2013).

to install solar panels: *CoEvolution Quarterly* (1974–1985), an offshoot of *The Whole Earth Catalog* (a counterculture golden-era mainstay) awarded this sun-challenged distinction to Forks in the late 1970s.

CHAPTER 8: In Search of the Holy Grail

from thirty years to ten: "A Big Bet on Small," *Economist Technology Quarterly*, December 6, 2014, 7.

so has it always been: Back in 1973, with fusion but thirty years away, there was another grail search under way, though not yet for storage. Unlike fusion, the goal of those Nixon era innovators we are hardly conscious of today because, in a rather unspectacular way, we found it. According to the papers of Southern California Edison the energy technology that fueled dreams in the early 1970s was a device that could directly convert "heat into electricity," eliminating "the need for shafts, turbines, and other rotating parts."

Given these conditions for success, solar panels and fuel cells, both extant but poor technologies at the time, were winning the optimists' race, for these two seemed best capable of charting a path around Carnot's theorem. Limits on the thermal efficiency of heat engines was by the late 1960s and early '70s crippling the long-solid electricity industry, so any means of making power without first burning fuel to heat water to make steam to turn a turbine was, effectively, a jewel-encrusted grail worthy of pursuit. At that time both solar cells and fuel cells were prohibitively expensive, barely viable technologies. But money and creativity and hope went, if sporadically, toward each. The half million American households that today have solar panels on their roofs are the beneficiaries of this once-ascendant vision of a more perfect energy source.

Even more noteworthy is that by 2014, machines designed to make power without steam had come to rule the market: 96.1 percent of all electrical power generation built in the United States that year was either natural gas, wind, or solar (53.3 percent wind and solar, 42.8 percent natural gas). Fuel cells, despite still being the object of flurries of activity and occasional press releases proclaiming success, still lag far behind their running mates in the "no steam for power" race to the top. This doesn't make them any less the repositories of hope, however, as K. R. Sridhar, the founder and driving force behind the Bloom Box, a recent addition to the struggling fuel cell market, makes clear. "We only wanted to solve the difficult problems, we didn't want to find the easy answer, we didn't want to find the first little thing that could work. We knew what we were after, we were after the holy grail and if we didn't get that we didn't care. That's the only thing we wanted and we went after it." K. R. Sridhar, "Boombox Energy Phenomenon." Interview available on YouTube.

as mercurial as a wind turbine: Balancing variable generation with other forms of variable generation can be disastrous because shifts in the sun and wind tend to be diurnal, which is to say, surface winds, which are driven by surface temperatures, have the unhelpful habit of slacking off at dusk. F. M. Mulder, "Implications of Diurnal and Seasonal Variations in Renewable Energy Generation for Large Scale Energy Storage," *Journal of Renewable and Sustainable Energy* 6, no. 3 (2014): 033105.

"the ensemble continuously produces beautiful music": Clay Stranger is a colleague of Amory Lovins. This a quote Lovins uses often to explain how storage might become part of a secure energy system that is always in motion. Not everyone agrees with him, but it is a compelling notion nevertheless. See especially Amory Lovins, "There Are Cheaper Ways to Keep the Lights on than Vast Electrical Storage," *Financial Times*, April, 13 2016. The comments section is also revealing. http://www.ft.com/intl/cms/s/0/a437955e-0098-11e6-99cb-83242733f755.html#axzz48YT31YG7.

fuels 98 percent of the local power plants: This statistic comes from an advertisement I saw at Ronald Reagan International Airport in January 2012 (it was produced by Friends of Coal, http://www.friendsofcoal.org/).

duration of the blackout: Edmund Conway, "World's Biggest Battery Switched on in Alaska," *Telegraph*, August 27, 2003, http://www.telegraph.co.uk/technology/3312118/Worlds-biggest-battery-switched-on-in-Alaska.html.

our national generating capacity: In 2013, data gathered from the *DOE Global Energy Storage Database*, accessed January 15, 2015, http://www.energystorageexchange.org.

regenerate an electric current: In Alabama, this is done with a natural gas combustion turbine.

there are effectively, none: The capacity is 336 MW in Colorado and 260 MW in Oklahoma, out of a total capacity of 18,341 MW nationally. "Licensed Pumped Storage Projects," *Federal Energy Regulation Commission*, April 1, 2015, http://www.ferc.gov/industries/hydropower/gen-info/licensing/pump-storage/licensed-projects.pdf.

in McIntosh, Alabama does: CAES stands for Compressed Air Energy Storage. There are only two storage facilities like this in the world: the McIntosh, Alabama plant, and one in Germany. Utah, Ohio, and Idaho also rest in part upon similar geologic structures, which are being considered for development now that effective storage has grown so critical to the good, future functioning of our grid.

almost twenty years of rechargeability: Flow batteries are a little bit like a regular battery turned inside out. Imagine the electrolyte being outside the terminals rather than surrounding them. As the electrolyte circulates it generates an electric charge across a membrane that separates the terminals. As it is depleted it can be topped off, rather like getting refills on one's coffee at a diner. For the moment the electron flow these batteries produce is not quite lustful enough to run much, and the very best chemistries involve using platinum (expensive) or vanadium (a rare earth chemical, found only in China, Russia, and South Africa and thus risky). Because of their potential for a long life—about 10,000 cycles or twenty years—these batteries nevertheless offer much promise. Specifics can be found at "BU-210b: How Does the Flow Battery Work?" *Battery University*, http://batteryuniversity.com/learn/article/bu_210b_flow_battery.

topography rather than technology: "So far, when Energy storage systems (ESSs) are integrated into conventional electric grids, special designed topologies and/or control for almost each particular case is required. This means costly design and debugging time of each individual component/control system every time the utility decides to add an energy storage system." Alaa Mohd et al., "Challenges in Integrating Distributed Energy Storage Systems into Future Smart Grid," in *IEEE International Symposium on Industrial Electronics* (IEEE, 2008), 1627–32.

0.8 percent of that states power: Alabama has the potential to produce 20 percent of its electricity from solar power; however, this is not being taken advantage of, in part because Alabama is only one of four states where homemade solar power is not purchased by the utility for redistribution. Their current fuel mix is coal 70.53 percent, nuclear 17.99 percent, gas & oil 9.55 percent, hydro 1.93 percent. Larry Clark, "Alabama Power, a Southern Company," at Grid Boot Camp, September 21, 2009.

"coal- and gas-fired power plants": Gillis (2015c).

"extend patterns in the present": Akhil Gupta, "An Anthropology of Electricity from the Global South," *Cultural Anthropology* 30, no. 4 (2015): 555–68.

doubled every year since 2005: Roxanne Palmer, "Solar Power Growing Pains: How Will Hawaii and Germany Cope with the Boom in Alternative Energy?" *International Business Times*, December 23, 2013, http://www.ibtimes.com /solar-power-growing-pains-how-will-hawaii-germany-cope-boom-alternative-energy 1518702.

in the country (after Arizona): Christian Roselund, "Arizona, Hawaii Lead the U.S. in Per-Capita Solar," *PV Magazine*, September 1, 2014, http://www. pv-magazine.com/news/details/beitrag/arizona--hawaii-lead-the-us-in-per-capita-solar_100016279/.

on the investment anywhere: "Top 10 States for Residential Solar—Fall 2014," *Solar Reviews*, accessed December 14, 2015, http://www.solarreviews.com /solar-power/top-states-for-solar-fall-2014-facts/.

makes its electricity from oil: So does Puerto Rico. Oil for electricity generation is a constant for island nations, island states, and island demistates.

times the national average: "U.S. Solar Market Trends 2013" (2014), 21.

solar power production: "Sunshine and Clouds," *Economist Technology Quarterly*, September 3, 2015.

produced from aggregate statistics: Alabama, Oklahoma, Arkansas, and Idaho are dead last in the 2015 solar power rankings (New York, Massachusetts, Connecticut, and Oregon take the first four spots). This list does not rank based upon installed solar power—California and Arizona don't even make the top ten—rather, it looks at how felicitous the regulatory environment is, at what the state and utilities are doing to make it affordable and easy to build solar, big and small, and link this back into the grid. "2015 United States Solar Power Ratings" http://www.solarpowerrocks.com/2015-solar-power-state-rankings/.

back into the public grid: "U.S. Solar Market Trends 2013" (IREC: Interstate Renewable Energy Council, July 2014), 15.

or on tropical islands: "Q2 2015 Solar Market Insight Fact Sheet," *Solar Energy Industries Association*, December 17, 2014, http://www.seia.org/sites /default/files/Q2%202015%20SMI%20Fact%20Sheet.pdf.

wannabe solar power producers: David Giles, "Blackout Insurance: Solar City Rooftops," *City Limits*, July 2, 2007, http://citylimits.org/2007/07/02 /blackout-insurance-solar-city-rooftops/.

Nevada and Utah markets: "Currently, Arizona is home to more than 20,000 residential solar customers, of which approximately 85 percent are solar lease households." Ian Clover, "Arizona to Impose New Tax on Solar Lease Customers," *PV Magazine*, May 7, 2014, http://www.pv-magazine.com /news/details/beitrag/arizona-to-impose-new-tax-on-solar-lease-customers-_ 100015000/.

"In 2007, only 10 percent of California homeowners were going solar through a solar panel leasing arrangement. The shift to over 75 percent solar leasing in 2012 is clearly significant." Zachary Shahan, "Solar Leasing Explosion In California (Chart)," *CleanTechnica*, December 9, 2013, http://cleantechnica. com/2013/12/09/solar-leasing-explosion-california-chart/. One company, Solar City, accounts for 32.5 percent of the market share residential solar installations in Arizona in 2012, and 17.2 percent in California. Andrew Krulewitz, "The Numbers Behind SolarCity's Success," *Greentech Media*, March 18, 2013, http://www.greentechmedia.com/articles/read/The-Numbers-Behind-SolarCitys-Success.

a whopping 307 in 2011: Lacey (2014).

underfunded distribution networks: Hertzog (2013).

"is not addressed in this way": Hertzog (2013)

"opposite of the traditional model": "All Change," *Economist*, January 17, 2015, 9.

in Western Europe it is already begun: Every European country is idiosyncratic. France, for example, is all about nuclear; this is part of the reason why France is working on fusion while Germany, which outlawed nuclear after Fukushima, is going toward renewables.

investing in a near future defection: Stephen Lacey, "This Is What the Utility Death Spiral Looks Like," *Greentech Media*, March 4, 2014, http://www .greentechmedia.com/articles/read/this-is-what-the-utility-death-spiral- looks-like.

"with a bunch of stranded assets": Lacey (2014).

they are largely inactive: According to Thomas Kuhn from the Edison Electric Institute, in the United States in 2010, many of our largest power plants were running at 10 to 15 percent of their potential; public presentation, Washington, D. C., September 23, 2009.

ways of generating revenue: For the two sides of the story see: Barbara Hollingsworth, "Report: Danger of Government-Created Solar Bubble Bursting When Subsidies Expire in 2016," *CNS News*, August 13, 2015, http://www.cnsnews.com/news/article/barbara-hollingsworth/report-danger- government-created-solar-bubble-bursting-when; and Jeff McMahon, "Solar's Future: Boom, Bust, Boom," *Forbes*, November 4, 2015, http:// www.forbes.com/sites/jeffmcmahon/2015/11/04/solars-future-boom- bust-boom/.

"pumped" hydro storage: Eduard R. Heindl, "Energy Storage for the Age of Renewables," *TEDxStuttgart*, March 11, 2013, https://www.youtube.com /watch?v=XF7mbEsEP04&index=9&list=PL9Xg-mFq2790Ok5edfcraCLA Ms5v4a-BI.

enough to make use of it: Energy Storage for Renewables Integration: Challenges and Successes Developing Solutions for Wind & Solar Assets, Navigant Research Webinar, January 13, 2015. http://www.navigantresearch.com /webinar/energy-storage-for-renewables-integration-3.

our common landscape: Richard F. Hirsh and Benjamin K. Sovacool, "Wind Turbines and Invisible Technology: Unarticulated Reasons for Local Opposition to Wind Energy," *Technology and Culture* 54, no. 4 (2013): 705–34.

almost exclusively in China: These elements are not for the most part rare, despite the appellation; they are difficult to depend on because of tariffs, and national borders, and unpredictable swings in currency rates. Tim Maughan, "The Dystopian Lake Filled by the World's Tech Lust," *BBC Future*, April 2, 2015, http://www.bbc.com/future/story/20150402-the-worst-place-on-earth.

both the market and the imagination: Sebastian Anthony, "At Long Last, New Lithium Battery Tech Actually Arrives on the Market (and Might Already Be in Your Smartphone)," *ExtremeTech*, January 10, 2014, http://www.extremetech.com/extreme/174477-at-long-last-new-lithium-battery-tech-actually-arrives-on-the-market-and-might-already-be-in-your-smartphone. See also "Battery Statistics," *Battery University*, accessed November 15, 2015, http://batteryuniversity.com/learn/article/battery_statistics.

someplace suspect, like Alberta: "About DLSC," *Drake Landing Solar Community*, accessed November 15, 2015, http://www.dlsc.ca/about.htm.

acid or even ceramic: While difficult to explain, batteries are quite simple to make. You will need a can of soda (any brand will do)—this is your electrolyte; a plastic, Styrofoam, or paper cup, which contains the process, like the wrapping on a battery; and a strip of copper that is slightly taller than the cup. Pour the soda in (drink what is left in the can), cut out a strip of the can that's roughly the same size as the copper strip, and stick each piece of metal in the cup, but don't let them touch. Voilà, a battery. If you connect the electrodes (the copper and aluminum strips) with a wire, about three quarters of a volt will stream between them. You could run a three-quarter-volt lightbulb with the thing, if such a bulb existed. This description is taken from "How to Make a Homemade Battery," *wikiHow*, accessed December 17, 2015, www.wikihow.com/Make-a-Homemade-Battery.

safer, and longer lasting: "Why Lithium Batteries Keep Catching Fire," *Economist*, January 27, 2014, http://www.economist.com/blogs/economist-explains/2014/01/economist-explains-19.

rainy days and long dark nights: Jake Richardson, "Tesla Powerwall Offered To Vermont Utility Customers . . . $0 Down," *CleanTechnica*, December 9, 2015,https://cleantechnica.com/2015/12/09/tesla-powerwall-offered-to-vermont-utility-customers-for-free/.

has always been their charm: Though somewhat scalable and somewhat portable, fuel cells suffer from being expensive and from the fact that they do need constant exposure to their fuel (they need to be plugged into natural gas pipelines, for example).

it seems, will be electric: Zack Kanter, "Autonomous Cars Will Destroy Millions of Jobs and Reshape U.S. Economy by 2025," *Quartz*, May 14, 2015, http:// www.nextgov.com/emerging-tech/2015/05/autonomous-cars-will-destroy-millions-jobs-and-reshape-us-economy-2025/112762/.

It sounds a little like Marxism: Karl Marx, *Critique of the Gotha Program* (Rockville, MD: Wildside Press, 2008 [1875]).

"when you are getting it fixed": quoted in Ryan Koronowski, "Why the U.S. Military Is Pursuing Energy Efficiency, Renewables and Net-Zero Energy Initiatives," *ThinkProgress*, April 4, 2013, http://thinkprogress.org/climate /2013/04/04/1749741/why-the-us-military-is-pursuing-energy-efficiency-renewables-and-net-zero-energy-initiatives/.

"dipping in the middle": *Economist* (January 17, 2015), 10.

out of the center at our grid: "How Much Energy Is Consumed in Residential and Commercial Buildings in the United States?" U.S. Energy Information Administration, accessed November 15, 2015, http://www.eia.gov/tools/faqs /faq.cfm?id=86&t=1.

incentives were different: Justin Gillis, "A Tricky Transition from Fossil Fuel: Denmark Aims for 100 Percent Renewable Energy," *New York Times*, November 10, 2014, http://www.nytimes.com/2014/11/11/science/earth /denmark-aims-for-100-percent-renewable-energy.html.

"Technology needs to save us": Gillis 2014.

(wealth of batteries to the grid's benefit): The popularity of electric and hybrid vehicles in Norway is largely the result of efficient, if often deemed excessive, subsidies and other benefits (like being able to use the HOV lane, even if you are driving alone or take ferries for free). Steve Hanley, "Electric Car Sales Surge in Norway during 2015," January 21, 2016, http://gas2 .org/2016/01/21/electric-car-sales-surge-in-norway-during-2015/.

municipal government—are electric: "Electric Vehicles and the Grid," *Navigant Research*, February 10, 2015, https://www.navigantresearch.com/webinar /electric-vehicles-and-the-grid.

just outside Hanover: The air force plans to expand the V2G demonstration to Joint Base Andrews, Maryland, and Joint Base McGuire-Dix-Lakehurst, New Jersey. The service will also continue to look for additional capabilities, such as utilizing used batteries as a form of on-base energy storage. "Air Force Tests First All-Electric Vehicle Fleet in California," U.S. Department of Energy, December 17, 2014, http://apps1.eere.energy.gov/news/news_ detail.cfm?news_id=21787.

designed in from the start: Michael d'Estries, "Operation Sustainability: U.S. Military Sets Ambitious Environmental Goals—Ecomagination," *Ecomagination*, January 30, 2012, http://www.ecomagination.com/operation-sustainability-us-military-sets-ambitious-environmental-goals.

"V2G demonstration in the world": "AF Tests First All-Electric Vehicle Fleet in California," *U.S. Air Force*, November 14, 2014, http://www.af.mil/News /ArticleDisplay/tabid/223/Article/554343/af-tests-first-all-electric-vehicle-fleet-in-california.aspx.

of our collective demand: Of the ten fastest-growing cities in the United States, nine are firmly in the AC zone (including five in Texas). "In Photos: The Fastest-Growing Cities In The U.S.," *Forbes*, accessed November 16, 2015, http://www.forbes.com/pictures/edgl45emig/no-1-raleigh-nc-metropolitan-statistical-area/.

"when there is no clear answer": Amelia Taylor-Hochberg, "Zoom In, Zoom Out: Hashim Sarkis, Dean of MIT's School of Architecture + Planning, on Archinect Sessions One-to-One #5," *Archinect*, accessed December 12, 2015, http://archinect.com/news/article/142833231/zoom-in-zoom-out-hashim-sarkis-dean-of-mit-s-school-of-architecture-planning-on-archinect-sessions-one-to-one-5.

CHAPTER 9: American Zeitgeist

"important qualities about them": A blogger going by the username "enkidu" notes that "seeky" is a linguistic alternative to the all-encompassing, "soul-destroying" definition of addiction advanced by the Partnership for a Drug-Free America. "Seeky," *everything2*, March 30, 2001, http://everything2. com/title/seeky.

something like "seeky": "Suppose you have a roof with a hole in it," the novelist Neal Stephenson explains, "that means it's a leaky roof. It's leaky all the time—even if it's not raining at the moment. But it's only lea*king* when it happens to be raining. In the same way, morphine-seeky means that you always have this tendency to look for morphine, even if you are not looking for it at the moment." Neal Stephenson, *Cryptonomicon* (New York: Avon Books, 1999), 373–74.

jet planes simply do not: In a widely publicized competition in July 2015, Airbus spent £14 million ($22 million) just to ensure the success of a single crossing of its electric plane over the English Channel. David Szondy, "Electric Aircraft Makes First English Channel Crossing," *Gizmag*, July 12, 2015, http://www.gizmag.com/first-electric-aircraft-cross-english-channel-airbus-cri-cri/38410/.

task-specific, and unobtrusive: Donald A. Norman, *The Invisible Computer: Why Good Products Can Fail, the Personal Computer Is So Complex, and Information Appliances Are the Solution* (Cambridge, MA: MIT Press, 1999), viii–ix.

fossil fuels for making power: Mitchell (2011) and Daniel Yergin, *The Prize: The Epic Quest for Oil, Money & Power* (New York: Simon and Schuster, 2011).

"filled with passengers": Fox-Penner (2014), xiii.

to bring down governments: Most famously, ordinary Egyptians and Tunisians used Facebook and Twitter to help coordinate their respective versions of the Arab Spring in 2011.

"don't want power lines.": Chris Kahn and Eric Tucker, "Easy Fix Eludes Power Outage Problems in U.S.," *Yahoo! Finance,* July 4, 2012, http://finance.yahoo.com/news/easy-fix-eludes-power-outage-problems-us-220940392.html.

"and out come fuels": David Rotman, "Praying for an Energy Miracle," *MIT Technology Review,* February 22, 2011, http://www.technologyreview.com/featuredstory/422836/praying-for-an-energy-miracle/.

rises to greet the grid: Hirsh and Sovacool (2013).

their bill for renewable power: Though for the most part, when it comes to the electric bill, very few people (about 10 percent) are willing to pay more than about 5 percent for "green" electricity. Michael Valocchi, IBM. Public presentation, September 22, 2009.

"beyond government mandates": She continues: "For those of you wondering where your energy comes from now via Excel [*sic*], here is a breakdown: The 2014 average mix of resources supplying Northern States Power customers includes coal (38.5%), nuclear (29.2%), natural gas (7.7%), wind (13.8%), hydro (7.7%), biomass (3.0%) and other (0.1%)." From Facebook, October 2015.

money for the cause: In keeping with this trend of feeling more comfortable with the less material something is, liquefied natural gas is often protested while natural gas in its gaseous form rarely is.

of the power they did in 1975: Roland Risser, "The Proof Is in the Pudding: How Refrigerator Standards Have Saved Consumers $Billions," *Energy.gov,* July 11, 2011, http://www.energy.gov/articles/proof-pudding-how-refrigerator-standards-have-saved-consumers-billions.

one with an Energy Star rating: Energy Star appliances allow one to save money on the electric bill over the life of the appliance, offsetting the higher purchase price, and they are a critical part of falling U.S. electricity consumption overall. Individuals, households, and appliances all use less electricity than they used to. So do manufactories, data centers, and large HVAC systems, because almost everything that relies on electricity has been redesigned, often numerous times, since the 1970s to use less of it.

take it one step further: The middle-aged still buy cars and houses, and still make choices about how to power those cars and houses, but the twentysomethings give every appearance of leaving even basic American materialism behind. They are undoubtedly the most wireless generation, but they also don't buy big stuff. The bigger the stuff, the less they buy of it. This is not only because they don't have the money for big-ticket items, but also because they are fundamentally uninterested in using the money they do have in ways that tie them to permanent debt or immobile assets. What to their parents looked like security (home ownership, car ownership, refrigerator ownership) to them feels like unjustifiable risk.

a cold spot built in?: Ikea already has a prototype for the non-fridge; the woven wall came from a conversation with a nineteen-year-old fiber arts student I know, who thinks bulbs are dumb and expensive.

"consistent time management": "NIST Framework and Roadmap for Smart Grid Interoperability Standards, Release 2.0" (National Institute of Standards and Technology, U.S. Department of Commerce, February 2012), http:// www.nist.gov/smartgrid/upload/NIST_Framework_Release_2-0_corr.pdf, pp. 79, 84, and 87, respectively.

One set of standards that most Americans likely assume are well enough in place have to do with cybersecurity. The more computerized the grid becomes the more vulnerable it becomes to hackers. It may be squirrels and tree branches today that do the most harm, but that's largely because the networks that will remake our grid over into a thinking machine are not yet well in place. See Koppel (2015). In 2001, a student in China wrote a research paper arguing that the entire Western Interconnection could be taken down by destroying just three substations. But just because hackers can take down our grid doesn't mean they want to. In response to a 2012 accusation by the *Daily Mail* that the hacking group Anonymous now had the capacity to "shut down the entire U.S. power grid" they said, "that's right, we're definitely taking down the power grid. We'll know we've succeeded when all the equipment we use to mount our campaign is rendered completely useless"(Gabriella Coleman, personal conversation, 2012). See also William Pentland, "Push Back: Utility Coalition Fights Federal Cyber Security Standards," *FierceEnergy*, September 24, 2015, http://www.fierceenergy .com/story/push-back-utility-coalition-fights-federal-cyber-security- standards/2015-09-24.

from renewables by 2030: Senate Bill 350, Clean Energy and Pollution Reduction Act of 2015.

has been installed since 2010: "U.S. Solar Market Trends 2013" (2014), 7 and 15.

in the past three years: According to IREC's "U.S. Solar Market Trends 2013," this number will change fast as more big solar comes online between 2015 and 2020. The California Valley Solar Ranch became operational in 2013 with generating capacity of 250 MW, while in 2014 alone, the Mojave Desert's Ivanpah Solar Electric Generating System (with a gross capacity of

392 MW), the Abengoa Mojave Solar Project (280 MW), and the Genesis Energy Solar Project (280 MW) were brought online. In 2015, Solar Star (579 MW) in Rosamund, Topaz Solar Farm (550 MW) in San Luis Obispo County, and the Desert Sunlight Solar Farm (550 MW) in the Mojave were each commissioned. Three more large-scale projects are either in the complaint phase or undergoing construction and are expected to be operational by 2020.

might come profitably together: Julia Pyper, "The Solar Industry Stands Divided Over California's 50 Percent Renewable Energy Target," *Greentech Media*, July 17, 2015, http://www.greentechmedia.com/articles/read/the-solar-industry-stands-divided-over-californias-future-renewable-energy. See also Beth Gardiner, "California Leads a Quiet Revolution," *New York Times*, October 5, 2015, http://www.nytimes.com/2015/10/06/business/energy-environment/california-leads-a-quiet-revolution.html, and Chris Megerian and Javier Panzar, "Gov. Brown Signs Climate Change Bill to Spur Renewable Energy, Efficiency Standards," *Los Angeles Times*, October 7, 2015, http://www.latimes.com/politics/la-pol-sac-jerry-brown-climate-change-renewable-energy-20151007-story.html.

"using less electricity more efficiently": Amory Lovins, "The Negawatt Revolution" in *Across the Board* XXVII vol. 9, September 1990, 21–22.

significantly more people: From Katherine Tweed, "U.S. Electricity Demand Flat Since 2007," *IEEE Spectrum*, February 6, 2015, http://spectrum.ieee.org/energywise/energy/environment/us-electricity-demand-flat-since-2007 and *Quadrennial Energy Review* 2015, DOE from a presentation at the Woodrow Wilson Center in D.C. on May 7, 2015. Nevertheless, it's worth considering "that the end-to-end infrastructure it takes to keep up with our Tweeting habit is responsible for more than 2,500 MWh per week of demand on the grid that simply did not exist before the application's advent." Massoud Amin, "Living in the Dark: Why the U.S. Needs to Upgrade the Grid," *Forbes*, July 11, 2012, http://www.forbes.com/sites/ciocentral/2012/07/11/living-in-the-dark-why-the-u-s-needs-to-upgrade-the-grid/.

and well-lit rooms: "Nebia Shower—Better Experience, 70 Percent Less Water" (2015).

simple as dimming the lights: Chris Mooney, "The Electricity Innovation so Controversial That It's Now before the Supreme Court," *Washington Post*, October 20, 2015, https://www.washingtonpost.com/news/energy-environment/wp/2015/10/20/the-electricity-innovation-so-controversial-that-its-now-before-the-supreme-court/.

running all over the place: "Several recent trends are creating an environment conducive to VPPs. These include the increasing penetration of smart meters and other smart grid technologies, growth in variable renewable generation, and emerging markets for ancillary services. However, challenges to commercial rollouts of VPPs remain, including the lack of reliance upon dynamic, real-time pricing and consumer pushback against the smart grid. The end goal for this market is the mixed asset VPP segment, as it brings distributed generation (DG) and demand response (DR) together to provide a synergistic sharing of grid resources." From "Virtual Power Plants" (*Navigant Research*, 2014), https://www.navigantresearch.com/research/virtual-power-plants.

where it might all be going: There was a recent essay in the *Atlantic* about how we mess up future predictions by focusing on the wrong things: how we get to work, for example, rather than what work is like. Thus, in the future as imagined in the 1950s, we all took jet packs to the office, but when we got there, there still weren't any women. Rose Eveleth, "Why Aren't There More Women Futurists?" *Atlantic*, July 31, 2015, http://www.theatlantic.com/technology/archive/2015/07/futurism-sexism-men/400097/.

at the same rate as a watt made: Brett Feldman, "All's Quiet on the DR Front, but a Storm Is Brewing," *Navigant Research Blog*, October 7, 2015, http://www.navigantresearch.com/blog/alls-quiet-on-the-dr-front-but-a-storm-is-brewing. See also Katherine Hamilton, "SPEER Releases Report on Benefits of Demand Response," *Advanced Energy Management Alliance*, October 29, 2015, http://aem-alliance.org/speer-releases-report-on-benefits-of-demand-response/.

"generators to increase supply": Mooney (2015). He goes on: "In particular, the objection before the Supreme Court is that this scheme of compensation gets FERC into regulating retail electricity markets, the ones that you and I are familiar with, because that's where we buy our own electricity from power providers. And FERC doesn't govern those—state public utility commissions do. What's complicated is that while demand response companies participate and bid into the wholesale markets—governed by FERC—their own clients are companies buying electricity on the retail markets, just like you and me (but generally on a much larger scale). In effect, demand-response blurs this distinction between the markets. And thus, before the Supreme Court, this has been framed as a battle over federalism, and whether FERC is going too far and getting into state territory."

"stimulate much-needed investment": "Selling It by the Negawatt," *Economist*, December 2, 2014, http://www.economist.com/news/business-and-finance/21635404-demand-response-industry-consolidating-selling-electricity-negawatt.

irregularities are highlighted: Donald Richie, *A Tractate on Japanese Aesthetics* (Berkeley, CA: Stone Bridge Press, 2007).

or a puddle of tar: Jacob Von Uexküll distinguishes between plants, which are immediately embedded in their habitat, and animals, which occupy a milieu (*Umwelt*). Perhaps von Uexküll's best-known example of this concept is the tick, whose milieu is constituted by a very limited number of factors. The tick climbs to the top of a branch or stalk and drops onto a passing animal, whose blood it then sucks. The tick has no eyes, the general sensitivity of its skin to sunlight alone orienting it in its upward climb. Its olfactory sense perceives a single odor: butyric acid, a secretion given off by the sebaceous follicles of all mammals. "When it senses a warm object below, it drops on its prey and searches out a patch of hair. It then pierces the host's soft skin and sucks its blood. The tick's milieu is made up of those elements that have meaning for it: sunlight, the smell of butyric acid, the tactile sense of mammalian heat, hair and soft skin, and the taste of blood. Its milieu is a closed world of elements, outside of which nothing else exists. Although it seems that animals all inhabit the same universe, each lives in a different, subjectively determined milieu" (58–59). In his discussion of Deleuze and Guattari's "animal music," as set forth in their "Of the Refrain" section of *A Thousand Plateaus*, Ronald Bogue reiterates the story of the tick, citing its source in "Uexküll's 1940 study *Bedeutungslehre* (Theory of Meaning)." Ronald Bogue, *Deleuze on Music, Painting and the Arts* (New York: Routledge, 2003).

AFTERWORD

I'll wonder if this is us too: Nathaniel Philbrick, *In the Heart of the Sea: The Tragedy of the Whaleship Essex* (New York: Penguin Books, 2001).

Index

access to electricity
  in early 1900s, 55–56
  as public right, 43, 46–47
  in rural areas, 55
Adams Street Station (Chicago), 64, 68, 71
aesthetics, 285–286
air-conditioning, 18, 67, 155, 170, 174, 176, 179, 237–238
Akins, Nick, 20
Alabama salt domes, 223
Albertsons grocery stores, 272–275
Alcoa Aluminum, 107
alternating current (AC)
  advantage over DC systems, 48–49
  big-city grids, 42
  as first step toward big grid, 48–50
  in networks, 51
  for Niagara Falls plant, 52–54
  physics of, 49
alternative energy, 281. See also renewable energy
American Electric Power, 130
American Telephone & Telegraph (AT&T), 58
American Tobacco, 58
Amin, Massoud, xiv, 176–177
Anaconda Copper, 58
analog meters, 150
Anderson, Brigadier, 217
animals, outages caused by, 211–212
antitrust laws, 60, 70
Apple, 190
Appleton, Wisconsin grid, 36–38, 56

arc lighting, 28, 30–35, 42, 44, 50, 54
Arizona, solar power in, 225, 230, 231
artificial lakes. See pumped hydro
Asmus, Peter, xiv

Bakersfield, California, 155–156
balancing authorities, 14
balancing load, 14, 175, 221. See also storing electricity
batteries
  components of, 240–243
  of electric cars, 243–251
  for energy storage, 221–224, 238–243
  flow, 224
  for military microgrids, 204–205
biological systems
  and the grid, 13–15
  longevity of, 197–198
blackouts. See power outages
Bogue, Ronald, 287
Bonneville Power Administration (BPA), 14
Boulder, Colorado. See SmartGridCity
brownouts. See power outages
Brush, Charles, 30, 34, 35, 44
Bryson, John, 101
Buffett, Warren, 58, 166
buried hydro storage, 236
Businessweek, xxviii
Byck, Peter, 203

California, 288
  Bakersfield protests, 155–156
  deregulation, 112–114

electric companies, 78–82
Enron's manipulation, 141–142
frontier electric technologies, 102–103
goals for renewables, xxii
pollution mitigation law, 92
power outages, xxvii, 190
renewable energy standard, 268–271
San Onofre power station, 116–118
small power producers, 98–99
solar power, 187, 224, 225, 230, 231, 269–271
treetop clearance, 121
wind power, 102–108
Canada, East Coast blackout and, 119
Carter, Jimmy, xxvi, 85–89, 91–92, 109
Casazza, Jack, 138
Cashman, Tyrone, 104–108, 187
Cataract Construction Company, 52
CenterPoint Energy, 151
central station grids, 36, 43–46, 269, 280
Chicago
    early electric companies, 50
    electricity use, 66
    private plants, 65
Chicago Edison, 63–69, 71–72
Chu, Stephen, 1–3, 10, 17–18
Cisco, 190
Citibank, xxviii
clean energy, xi. *See also* renewable energy
climate change, xix, xxvii–xxviii, 225
coal, xvi, xviii–xix
coal-burning plants, 222
    carbon dioxide from, 20
    replacements for, 118
    shrinking number of, 117
    as stock resource, 9
    switch to oil-burning plants, 75
cogeneration, 94, 96–97, 102–103, 108–109
Colorado. *See* SmartGridCity
Columbia River Gorge wind farms, 3, 5, 10–17
common good, power as, 43, 46–47, 77
Commonwealth Edison, 84
competition, 27, 41–43. *See also* Public Utilities Regulatory Policies Act (PURPA), section 210
    among service provider plans, 158
    and electricity as public good, 77
    and Energy Policy Act, 120, 142
    opening door to, 109
    from private plants, 43

and rise of monopolies, 57–59 (*See also* monopolies)
    in wireless market, 284
compressed-air storage, 223–224, 236
computer platforms, 261–261
computer-related power outages, 131–134
computers, xi–xii, 171
concentrating solar towers, 224–225
ConEd, 166
Connecticut, xxviii
conservation, 85–87
    capacity to deploy, 279
    and control by customers, 157–158
    environmental movement, 75–76
    and reform of energy system, 262, 267, 271
Consolidated Edison, 174
consolidation of electricity services, 59–60.
    See also monopolies
consumer culture, 47–48, 86
copper theft, 212–213
cultural systems
    consumer culture, 47–48, 86
    and early private grids, 42–43
    and the grid, 13–15
    mass market, 47–48
culture of electricity making, 4

dams, 12–13, 19, 28, 104, 117
Davis, Gray, 114
Davis-Besse Nuclear Power Station (Ohio), 115–116, 144
demand-response (DR), 276–277
demand-side reform, 149–183. *See also* energy security; human interactions with grid
    and ability of utilities to meet demand, 176–179
    to control peak demand, 152–154, 178–179
    to control peak load, 173–174
    to control revenue streams, 152
    and grid defection, 167–168
    and Internet of Things, 170
    microgrids, 167–168
    nanogrids, 167–172
    net metering, 158–159
    new attention to, 152
    and predictability of demand, 174–175
    smart grid, 159–165, 171–172
    smart meters, 149–152, 154–158, 165, 172–174, 180–181, 276
Denmark, 107–108, 247–248

Department of Energy, 88
deregulation, 21, 110–114
Detroit blackout (2014), 196
Detroit Edison, 69
diesel generators, 188–189, 203, 213, 214, 221
direct current (DC), 49
  advantage of AC over, 48–49
  big-city grids, 41
  comeback of, 29
  limitations of, 34
  in networks, 50–51
  and Niagara Falls plant, 52–53
  physics of, 49
  for privately owned power systems, 29, 36, 37, 45
  rotary converters, 52
  for traction companies, 54
disasters, 115–148
  and attempts to control systems, 135
  computer-related, 131–134
  East Coast blackout (2003), 119–126, 128–138
  and Energy Policy Act, 138–140, 142, 145, 147
  and instability of the grid, 138
  and movement of electricity, 126–128
  nuclear plant vulnerabilities, 115–117
  political and economic factors in, 137–145
  power plant failures, 115–116
  Swiss Cheese Model of Industrial Accidents, 135–136
  tree-related, 119–125
  from unpredictable renewables generation, 2–3
distribution networks/systems, 232
  automatic shut-offs in, 125
  in early grid, 54
  power outages starting on, 145
  and replacement of aging plants, 118
  with rooftop solar, 228–232
  separation of generation from, 120
  separation of transmission systems from, 138
  and tree-related power outages, 119–125
  wireless, 282–283
Duke, J. B., 58
Duke Energy, 166
Dupee, Steve, 131
DuPont, 58
Durkin, John, 96
dynamos, 30, 32, 34, 35, 37, 41

East Coast blackout (1965), xv, 196
East Coast blackout (2003), xv, 119–126, 128–138
Eastern Interconnection, 141, 147
economic factors, in disasters, 137–145
economic impact, of power outages, xiv–xv
Economist, 233, 246
Edison, Thomas Alva, 31, 32, 34–37, 42, 44, 50, 59, 61–63
Edison Electric Institute, xxviii
Edison General Electric Company, 44, 47
efficiency, 279
  energy, 85–86
  lighting, 265
  as means of generating less power, 276
  power plant, xiv, 71–75, 98
  and reform of energy system, 262, 267, 269
electric cars, electricity storage and, 243–251
electric companies. *See also* utilities companies; *individual companies*
  benefits of regulation for, 77
  business models of, 81, 86
  in California, 78–81
  deregulation of, 110–114
  and early AC vs. DC use, 50–51
  failure of, 182–183
  for-profit, investor-owned, 78
  grow-and-build strategy of, 72–75, 77
  as monopolies, 60–63, 69–71, 110
  as monopsonies, 93–104
  small, 78
electricity, 5–6
  AC, 48–50
  access to (*See* access to electricity)
  addiction to, 255–257
  from cogeneration, 97
  as commodity, 138–140
  as common/public good, 43, 46–47, 77
  DC, 29, 48–51, 49
  dependence on, xi–xii, 291–292
  drop in consumption, 86, 91, 96, 151, 274, 275
  first-generation uses of, 29–30
  flow of, 6–7
  instantaneous power at a distance from, 25–27
  knowledge about, 25–26
  making, xxiii
  movement of, 126–128
  phase converters, 54
  predictability of, 141–142

price of (*See* price of electricity)
rotary converters, 52, 54
storing (*See* storing electricity)
transition from serial to parallel circuitry,
    31–36
voltage, 24, 38–41, 48–49
Electric Power Supply Association (ERSA),
    278
Enercon, 15
Energy Cloud, 280
energy crisis (1970s), 85
energy efficiency, 85–86
Energy Futures Holding Corporation, 166,
    183
energy independence, xxviii
Energy Information Administration (EIA),
    xviii–xix
Energy Policy Act (1992), 21–22, 119–120,
    142
    and energy trading, 138–140
    and power quality, 145
    as reregulation, 111
    results of, 147
energy security, 185–218
    fuel for military isolated grids, 202–209
    local independent energy systems,
        185–192
    microgrids, 199–204
    and non-storm-related outages, 210–213
    resiliency, 191–204
energy trading, 138–140, 142
England, 288
Enron, 113, 141–142, 182–183
environmentalism, 90–92
environmental movement, 75–76
Europe, 233

Facebook, 190, 261
Fairfax, Stephen, 164
Faraday, Michael, 29
Federal Energy Regulatory Commission
    (FERC), 100, 120, 278, 279
FirstEnergy, 119, 166
    Davis-Besse upgrades by, 144
    East Coast blackout, 119–126, 128–135,
        137–138
Fisk Street Station (Chicago), 72, 84
Fleishman, Glenn, 173, 176
flow batteries, 224
fluorescent lighting, 61
Forks, Washington, 214
for-profit monopolies, 81
Foshay, Wilber, 69

fossil fuel power plants, 74
frontier electric technologies, 97, 102–103
Fukushima nuclear reactors (Japan), 117

General Electric (GE), 47–48, 63
generation. *See also* power plants
    and Energy Policy Act, 141, 143
    and 2015 Paris climate change talks,
        225
    from pumped hydro, 223
    from renewables, 234
    separation of distribution from, 120
generators
    diesel, 188–189, 203, 213, 214, 221
    dynamos, 30, 32, 34, 35, 37, 41
    for U.S. military, 203, 206–208
Germany, xiv, 233–234, 236, 288
Gipe, Paul, 106
Google, xxviii, 190, 217
Gorguinpour, Dr., 245, 251
green energy, xvi–xviii, 262–263. *See also*
    renewable energy
Green Mountain Power, 243
the grid, xii–xiii. *See also specific topics*
    command and control structure of, 22
    components of, xiii, xiv
    and cultural systems, xviii, 13–15
    current capabilities of, 15–17
    current efforts toward fixing, xxix–xxx
    functioning of, xxii–xxv
    and goals for renewables, xxii
    history of (*See* history of the grid)
    as just-in-time system, 7
    local nature of, xxi
    microgrids, xxviii
    networks of business interests, geo-
        political stakeholders, and legal
        structures of, xix–xx
    overhauling (*See* reforming the grid)
    power outages, xii, xiv–xvi
    solutions to limitations of, 17–18
    technological, biological, and cultural
        systems at work in, 13–15
    three sections of, xiv
    transforming, xxiv–xxv
    and transition to green energy, xvi–xviii
grid defection, 167–168
    Americans' reluctance for, 233–235
    developments facilitating, 281–282
    in early 2000s, 114
    and utility death spiral, 233–235
grow-and-build strategy, 72–75, 77
Gupta, Akhil, 226

Haney, John, 56
hard path/hardening, 193–194
Harrison Street Station (Chicago), 71–72
Hawaii
  goals for renewables in, xxii
  home solar systems in, xxi, 20–21, 226,
    227, 230
Hirsh, Richard, 57–58, 70, 92–93
history of the grid, xxv–xxvii
  absolute power of utilities in 1970s,
    92–93
  AC electrical systems, 48–51
  central station grids, 36, 43–46
  chaotic infrastructure, 51–52
  Chicago Edison, 63–69, 71–72
  cogeneration plants, 96–97, 102–103,
    108–109
  competition, 41–43, 109
  consolidation of electricity services,
    59–60
  DC systems, 29, 49–51
  deregulation, 110–114
  early 1900s access to electricity, 55–56
  early electricity as local affair, 24
  early product development, 27–30
  early technology development, 26–27
  electric power monopolies, 69–71, 110
  energy crisis, 85
  environmentalism, 75–76, 90–92
  first municipal grid, 36–38
  grow-and-build strategy, 72–75, 77
  and knowledge about electricity, 25–26
  lighting, 28–38
  municipal vs. for-profit-company
    monopolies, 81
  National Energy Act, 86–88
  Niagara Falls power plant, 52–54, 56
  PG&E and local community governance of
    electric power, 79–81
  and power as a common good, 47
  private grids, 42–47, 51
  private plants, 28–29, 36, 37, 43–47, 60,
    65
  proliferation of wires, 41, 42, 51
  PURPA, 87, 93–104
  regulation, 77–79
  renewable power projects, 97–99,
    102–109
  rise of mass market/consumer culture,
    47–48
  rise of monopolies, 57–63
  rotary converters, 52
  small power innovators, 96–103
    strategies for preserving monopolies,
      80–81
    switch to oil-burning power plants, 75
    technological improvements and plant
      efficiency, 72–75
    technology dependence by 1970s, 88–89
    Three Mile Island, 91
    transformers, 49
    transition from serial to parallel circuitry,
      31–36
    in urban areas, 40–41
    utilities as monopsonies, 86–88, 93–104
    voltage, 38–41, 48–49
    water-powered dynamos, 37
    wind power in California, 102–108
home appliances, 47–48, 55, 56, 67, 82, 174,
    265–266
home/rooftop solar systems, xvii–xviii,
    226–232
  battery back-ups for, 221
  in California, 269–271
  and crisis of infrastructural management,
    20–21
  in Hawaii, xxi, 20–21, 226, 227
  as money-making schemes, 281
human interactions with grid, 255–260,
    272–289
  addiction to electricity, 255–257
  cost of retrofitting, 274–275
  and imperfect things, 285–287
  and infrastructure visibility, 258, 263–264
  and need for grid overhaul, 259–260
  negawatts, 272–277
  and preference for common grid, 285
  surcharge for renewables, 264–265
  virtual power plants, 277–281
  wireless systems, 282–285
Hunter, Don, 131
Hurricane Irene, 196
hydropower, 94
  buried hydro, 236
  in California, 104
  dams, 12–13, 19, 28, 104, 117
  for early small grids, 37
  for lighting Sierra Nevada gold mines, 30
  in Pacific Northwest, 10
  as percentage of renewable energy, 19
  pumped hydro, 223, 225, 236
  shrinking number dams, 117

Iberdrola, 14–15
Ice Bear energy storage system, 238
iceboxes, 238

Ikea, 262, 282
incandescent lighting, 28, 31, 34, 36–40, 51, 265
infrastructure. *See also* the grid; *individual parts of infrastructure*
 early chaos of, 51–52
 early proliferation of wires, 41, 42, 51
 lagging changes in, 22–23
 line loads after Energy Policy Act, 140–141
 scale of, xxviii
 virtual power plants, 277–281
 visibility of, 54, 237–238, 258, 263–264, 282
 working with government on, 70
instability of the grid, 138
Insull, Samuel, 61–69, 71, 72, 80, 82–84, 92, 178, 281
intelligent grid, 287–288
Interim Standard Offer #4 (ISO₄), 102–105, 107
Internet, 112, 142
Internet of Things, 170, 289
interoperability standards, 268–269
Iowa, wind power in, xxi
islanding, 130, 198, 199, 251
Italy, xiv

Japan, xiv, 285, 286
Jenkins, Phillip, 203
Johansson, Bob, 189–190, 215, 216
Jones, Chuck, 137, 138

Keates, John, 253
Kelvin, Lord, 53
Klein, Maury, 62–63
Korea, xiv
Krzanich, Brian, 284

Las Vegas county, Nevada, 156–157
legacy technologies, 262–272
Lerner, Eric J., 127
Lewis, Nate, 263
lighting
 arc, 28, 30–35, 42, 44, 50, 54
 early technologies/grids for, 28–38
 early urban grids, 41, 42
 efficiency of, 265
 fluorescent, 61
 incandescent, 28, 31, 34, 36–40, 51, 265
 predictability in use of, 174
 transition from serial to parallel circuitry, 31–36

lithion-ion batteries, 240–243
load(s)
 after Energy Policy Act, 140–141
 balancing, 14, 175, 221
 off-peak power, 67–68
 peak (*See* peak load)
 variability of, 66–67
local community governance of electric power, 79–81
local independent energy systems, 185–192
Lockheed Martin, 219, 220
Los Angeles Air Force Base, 250–251
Lovins, Amory, xxvi, 194–198, 205, 217, 271, 273
Lovins, Hunter, xxvi, 194–198, 217, 271

macrogrid, 199. *See also* the grid
Maine
 goals for renewables in, xxii
 smart meters in, 167
Mainzer, Elliot, 16–17
Marin Clean Energy, 80
Marin County, California, 79–81
Massachusetts, 229
mass market, 47–48
Medicare, 104
microgrids, xxviii, 167–168. *See also* private grids
 in early days, 24
 for energy security, 199–204, 215–216
 military, 202–209
 for resiliency, 198
microprocessors, 287
Midwest Independent Transmission System (MISO), 131
Milburn, Ski, 163
military installations, xxix
 microgrids of, xxviii, 202–209
 storing electricity at, 250–251
Mississippi salt domes, 223
mobile power station, 208
monopolies, 69–71, 110. *See also individual companies*
 guaranteed profits of, 74
 municipal utilities as guaranteed profits of, 79
 municipal vs. for-profit, 81
 and National Energy Act, 86–88
 rise of, 57–63
 strategies for preserving, 80–81
monopsonies, 86–88, 93–104, 112
Morgan, J. P., 58, 69
Motyka, Theodore, 204

Muha, Adam, 123
municipal grids, 59–60
   in California, 78–79
   in early 1900s, 60
   first, 36–38, 44
   governance of, 59
   as monopolies, 79
municipal monopolies, 81
Munson, Richard, 68–69

nanogrids, 167–172, 200. *See also* private
   grids
NASDAQ power outages, xv, 212
National Energy Act (1978), 86–88, 111
National Energy Plan, 88
National Renewable Energy Laboratory
   (NREL), 18–19, 88, 109
national security, xxvi
natural gas, xix
natural gas plants, 9
natural gas turbines, 221
negawatts, 272–277
Neri, Joseph, 30
net metering, 20, 158–159
Nevada
   Las Vegas smart meter radiation, 156–157
   solar power in, 224, 225, 231
New Hampshire, 229
New Jersey Transit, 201, 216, 217
New York
   six-cent law in, 101–102
   Superstorm Sandy, 191–192, 198, 199,
      215
New York City
   blackout of 1977, xv, xxviii, 196
   peak demand reduction in, 179
   Pearl Street Station grid, 33, 34, 44, 50, 56
   private grid wires in, 41, 42
New York University (NYU), 201
Niagara Falls power plant, 27, 52–54, 56
Norman, Donald, 257
nuclear fusion, 219–220
nuclear power plants, xix
   Davis-Besse Nuclear Power Station,
      115–116
   deterioration of, 116–118
   fall of Vermont nuclear plant, xxvii
   protests of, 77, 91
   reliability of, 174
   as stock resource, 9
   Three Mile Island, 91
   Vermont Yankee power plant, 116
   vulnerabilities of, 115–117

Obama energy plan, 17–20
office towers, batteries that look like,
   239–240
off-the-grid, 167, 213–214
Ohio, East Coast blackout and, 119–126
oil-burning power plants, 75
Open Source Initiative, 180
Oregon, wind power in, xxi
Ozarks blackout (1987), xv

Pacific Northwest
   hydroelectric infrastructure in, 10
   power outages, xxvii, 185–191, 214–215
parallel circuits, 31–36
peak demand. *See also* storing electricity
   demand-side reforms to control,
      152–154, 178–179
   power plants reserved for, 152–153,
      177–178
peak load, 8
   demand-side reforms to control, 173–174
   increase in, 175–176
   in smart grids, 171–172
Pearl Street Station (New York), 33, 34, 44,
   45, 50, 56
Pepco, 179, 180
Peterson, Bud and Val, 168–172
PG&E, 79–81, 121, 155, 156, 166, 182
phase converters, 54
Pickel, Fred, 112
platforms (computer), 261–261
political factors, in disasters, 137–145
pollution, 89, 90
Porter, Kevin, 103
Post, Gene, 131
potentiality, 39. *See also* voltage
Powerall battery system, 242–243
power outages, xxvii. *See also individual
   outages*
   causes of, xv–xvi
   computer-related, 131–134
   consumers' attitudes toward, 118
   cost of, 205
   on distribution systems, 145
   duration of, xiv
   impact of, xiv–xv, 291–292
   increase in, xiv
   and movement of electricity, 126–128
   non-storm-related, 210–213
   and resiliency, 191–204
   rolling blackouts, 113
   storm-related, 185–192, 214–215
   in transmission systems, 232

tree-related, 119–125
and underfunded distribution networks,
    232
power plants, 28. *See also specific plants and
    types of plants*
efficiency of, xiv, 71–75, 98
environmental regulations and cost of
    building, 76
failures of, 115–116
first large-scale, 52–54, 56
grow-and-build strategy, 72–75, 77
money owed on, 234
for renewable sources, xx–xxi
reserved for peak demand, 152–153,
    177–178
virtual, 277–281
predictability
of demand, 174–175
of electricity, 141–142
of generation from renewables, 1–10
price of electricity
in 1970s, 72
and amount of electricity used, 81–82
and conservation of power, 272
and energy trading, 139
and environmental movement, 76
fixed, 71
in Germany, 233
in Hawaii, 226
for industry vs. residential/commercial
    customers, 68
and money made by company, 68
in monopsonies, 92, 99–101
for off-peak vs. peak power, 67
and oil embargo of 1973, 75
promotional rates, 95–96
and rooftop solar systems, 228–231
surcharge for renewables, 264–265
tiered rate structures, 92
and use of smart meters, 154–156
and utility death spiral, 233–235
private grids, 42–47, 51. *See also* home/
    rooftop solar systems; microgrids;
    nanogrids
private plants, 28–29, 36, 37, 43–47, 60, 65,
    233–234
product development
for early grid, 27–30
lighting, 28–38
promotional rates, 95–96
public good, power as, 43, 46–47, 77
Public Utilities Regulatory Policies Act
    (PURPA), section 210, 87, 93–104, 114

Public Utility Holding Company Act (1935),
    60
pumped hydro, 223, 225, 236

radiation
    from smart meters, 156–157
    from wireless transmission, 284
Rancho Cucamonga, 78–79
Reagan, Ronald, 106
rebates, for home solar, 229
reforming the grid, 23, 255–289. *See also*
    demand-side reform
    dealing with combined interests of
        players, 260–262
    dealing with legacy technologies, 262–272
    human interactions with grid, 255–260,
        272–289
    kinds of problems in, 260
    virtual power plants, 277–281
regulation, 70, 74, 77–79
    and deregulation, 21, 110–114
    environmental, 76
    as monopolies, 94
    as monopsonies, 95
    by utilities, 92
renewable energy, 262–263. *See also*
        sustainable energy; *specific types, e.g.:*
        wind power
    California standard for, 268–271
    consumer surcharge for, 264–265
    and current capabilities of the grid,
        15–17
    current capacity for integrating, 221
    generation from, 234
    goals for, xxii
    for living off-the-grid, 213–214
    new generation from, 109
    in Obama energy plan, 17–20
    overproduction of, 10–13
    under PURPA, 97–99, 102–104
    shift to, xx–xxi
    subsidies for, 14–15, 231
    tax credits for, 104–105
    variable generation from, 1–10
resiliency, 191–204
REW, 234
Robert Moses Niagara Power Plant, 56
Rockefeller, John D., 57, 58
rotary converters, 52, 54
Rural Electrification Act (1936; REA), 55, 81

salt domes, 223–224
salt towers, 224–225

San Diego Gas & Electric, 272
San Francisco
    early arc lighting in, 30–31
    first grid in, 27
San Onofre power station, 116–118
serial circuits, 31–36
Silicon Valley substation, 211
sinks, 128–130
small power producers, 96–103, 269–271.
    See also home/rooftop solar systems
smart grid, 159–165, 171–172
SmartGridCity, 159, 161–166, 168–172,
    181–182
smart meters, 149–152, 154–158, 165,
    172–174, 180–181, 276
smart phones, xiii
smart thermostats, 162, 163
Snickey, Jerry, 131, 132
soft energy technologies, 194–198
solar duck conundrum, 244, 246
solar power, xvii. *See also* home/rooftop
    solar systems
    concentrating solar towers, 224–225
    condensed solar plants, 21
    for microgrids, 187
    overproduction of, 10–13
    price of solar panels, 230
    recent growth in, 19
    solar duck conundrum, 244, 246
    solar trough plants, 225
    storing, 221, 224–226, 243, 245–246
    Walmart solar panels, 267
solar trough plants, 225, 246
Soljačić, Martin, 284
Southern California Edison, 69, 78, 81–82,
    246
South Oakes Hospital, Long Island, 199
Southwest blackout (2011), 196
Soviet Union, 43, 93
Standard Oil Trust, 57, 58
standards for interoperability, 268–269
steam plants, 73–74
"stock resources," xvi, 9
storing electricity, 5, 219–253
    batteries, 221–223, 238–243
    buried hydro, 236
    compressed-air storage, 223–224, 236
    concentrating solar towers, 224–225
    and disposing of surplus power, 10
    in electric cars, 243–251
    at grid scale, 222–223
    and home solar systems, 226–232
    in icebox, 238

"invisibility" of infrastructure for,
    237–238
    at military bases, 250–251
    non-battery ideas for, 235–237
    pumped hydro (artificial lakes), 223, 225,
    236
    salt domes, 223–224
    solar trough plants, 225, 246
    and utility death spiral, 233–235
Strategic Petroleum Reserve, 88
Straw, Daniel, 186, 216–218
Straw, Sylvie, 185–189, 216–218
subsidies, for renewables, 14–15, 231
substations, xiii
SUNY Stony Brook microgrid, 199
Superstorm Sandy, 188, 191–192, 198, 199,
    215
sustainable energy, xi, xvii–xviii, 262–263.
    See also renewable energy
Swisher, Randall, 103

Tactical Alternating Current System
    (TACS), 207–208
Tactical Garbage to Energy Refinery
    (TGER), 208–209
Taormina, Mr., 149–151
Taormina, Thelma, 149–151
tax credits
    for renewable energy generation, 104–105
    for rooftop solar systems, 229, 269
technologies. *See also individual technologies*
    adoption of, 55
    Americans' lack of knowledge about, 89–90
    early development of, 26–27, 29
    and the grid, 13–15
    growing dependence on, by 1970s,
    88–89
    Internet, 112
    legacy, 262–272
    plant efficiency and improvements in,
    72–75
    SmartGridCity, 159, 161–166, 168–172,
    181–182
    telecommunications, 159–160
telecommunications industry, 159–160,
    172
Tennessee Valley Authority (TVA), 117
Tesla, Nikola, 31, 50, 282–284, 289
Tesla Motors, 242–243
Texas
    blackouts in, xxvii
    renewable power generation in, 109
    wind power in, xxi–xxii, 15, 108

Three Mile Island, 91
transformers, xiii, 49
transmission systems, 232–233
    automatic shut-offs in, 125
    for early grids, 26–27
    and grid stability, 9–10
    for Niagara Falls plant, 52–53
    during peak demand, 153
    separation of distribution systems from,
        138
    upkeep and new construction of,
        144–145
    wireless, 282–285
Transportable Hybrid Electric Power Station
    (THEPS), 208–209
tree-related power outages, 119–125
TV pickup, 153–154
Twitter, 261

Uber, 261
United Corporation, 69
United Kingdom (UK), 153–154
universal system, 54, 268
universities, central air facilities in, 54
University of California, San Diego,
    201–202
urban areas, 40–41. *See also individual cities*
U.S. Steel, 58
Utah, solar power in, 231
utilities companies, xxvi. *See also* electric
        companies
    ability to meet demand, 176–179
    absolute power of, in 1970s, 92–93
    and changes in the grid, 233
    death spiral for, 88, 166, 181, 233–235
    deregulation, 111–112
    early central station grids, 36, 43–46
    effect of PURPA on, 99
    FERC sued by, 279
    and home solar systems, 20
    information for, 142–143
    as monopolies, 57–63, 69–71, 80–81
    as monopsonies, 93–104, 112
    profits of, 74, 143
    regulation of, 70, 77–79, 114
    and rise of mass market/consumer
        culture, 47–48

variable generation, 1–10
    and Energy Policy Act, 21–22
    and PURPA, 99
    and replacement of aging plants, 118
vars (reactive power), 145–146
vehicle-to-grid storage systems, 243–251
Vermont
    fall of nuclear plant in, xxvii, 116
    goals for renewables in, xxii
    solar power in, xxi, 229
Vermont Yankee power plant, 116
Virginia Power, 102
virtual power plants, 277–281
visibility of infrastructure, 54, 237–238, 258,
        263–264, 282
Volta, Allesandro, 29, 240
voltage, 38–41
    in AC systems, 48–49
    in early electrical systems, 24, 37–38

Walmart, 267
Warwick, Mike, 163
Washington, D.C., 179
WE Energy, 56
Western Doughnut, 11, 46
Western Interconnection, 22, 146, 251
Westinghouse, George, 42, 44, 50
White House power outages, xxvii
wind power, xvii
    balancing, xxi–xxii
    in California, 102–108
    overproduction of, 10–13
    recent growth in, 19
    in Texas, xxi–xxii, 15, 108
    turbines, 106–108
    unpredictability of, 3, 5, 8–9
wireless, 160–163, 249, 263, 264, 282–285.
        See also smart meters
Wisconsin Energy corporation, 56
wise grid, 268, 288
Wollenberg, Bruce, 263
WorldCom, 172

Xcel, 159, 161–166, 168–172, 179–182,
        264

Yahoo, 190

A Note on the Author

Gretchen Bakke holds a Ph.D. from the University of Chicago in cultural anthropology. She spent the first half of her career researching failing nations, including the Soviet Union and the former Yugoslavia, before turning her attentions to failing infrastructural systems. She is a former fellow in Wesleyan University's Science in Society program and is currently an assistant professor of anthropology at McGill University. Born in Portland, Oregon, Bakke lives in Montreal but calls D.C. home.